Basic Metrology for
ISO 9000 Certification

Basic Metrology for ISO 9000 Certification

G.M.S. de Silva

Oxford Auckland Boston Johannesburg Melbourne New Delhi

Butterworth-Heinemann
Linacre House, Jordan Hill, Oxford OX2 8DP
225 Wildwood Avenue, Woburn, MA 01801-2041
A division of Reed Educational and Professional Publishing Ltd

 A member of the Reed Elsevier plc group

First published 2002

British Library Cataloguing in Publication Data
A catalogue record for this book is available from the British Library

Library of Congress Cataloguing in Publication Data
A catalogue record for this book is available from the Library of Congress

ISBN 0 7506 5165 2

For information on all Butterworth-Heinemann
publications visit our website at www.bh.com

Typeset at Replika Press Pvt Ltd, Delhi 110 040, India
Printed and bound in Great Britain by Biddles Ltd, *www.biddles.co.uk*

Contents

Foreword

Metrology is often considered to be simply a field of science and technology concentrating on the measurement and accuracy of things we make. However, much of the historical development of technology suggests this chicken and egg situation is reversed, in that evolving the ability to measure qualities more accurately allows us to develop more reliable ways of manufacturing things, as we now have a means of controlling quality.

The recent rise in nano-technological advances is a beacon to this premise, which was recognized in an earlier era of technological development, that might be called the micro-technology phase associated with the industrial revolution. In particular by prominent people of that era, such as one of the founders of modern metrology, Joseph Whitworth, who in his Presidential Address to The Institution of Mechanical Engineers, in London in 1856, said the following;

"I would next call your attention to the vast importance of attending to the two great elements of constructive mechanics, namely, the true plane and the power of measurement",

and

"I hope the members of this institution will join me in doing what we can with reference to these two important subjects – correct measurement and its corollary proper graduation of size. The want for more correct measurement seems to pervade everything."

Little has changed in our needs in this respect, and the importance of metrology to the advancement of science and technology remains, albeit on improved levels of accuracy.

The author, Swinton de Silva, has worked for the majority of his working life, a period covering several decades, in this field, and the impressive breadth of his metrology knowledge is reflected in the very broad ranging contents of this book. As well as representing an extremely valuable contribution to the literature in the metrology field, this book also represents an essential guide to those involved with ISO 9000 Certification.

Dr R S Sayles
Reader in Mechanical Engineering
Imperial College of Science, Technology and Medicine
London, UK

Preface

Test and measuring instruments are used extensively in the modern manufacturing and processing organizations. The accuracy of the measurements made by these instruments has a direct impact on the quality of the product or service provided by the organization. The recent developments in the field of certified quality and environmental management systems, namely registration to ISO 9000 and ISO 14000 standards, require that test and measurement equipment be periodically calibrated using measurement standards traceable to the international measurement system. In addition there is also the necessity for test and calibration laboratories to be accredited by a third party certification body in accordance with the international standard ISO/IEC 17025 (previously ISO/IEC Guide 25).

Although a number of books are available describing specific measurement fields such as temperature and pressure, books covering a number of important measurement fields are few. This book intends to fill this gap. The book is primarily aimed at persons working in industry whose duties are related to calibration and maintenance of test and measurement equipment. Students reading for bachelor's degrees or diplomas in the fields of electrical, mechanical and production engineering and related technology-based courses can also use it as an introduction to metrology.

The book is an introduction to fundamental measurement principles and practical techniques used in the calibration of test and measuring equipment belonging to seven measurement fields, namely length, angle, mass, temperature, pressure, force and electrical metrology. Fundamental concepts of measurement and calculation of measurement uncertainties are also dealt with.

G M S de Silva

Acknowledgements

The author wishes to acknowledge gratefully the many persons who assisted him during the course of the preparation of the book. In particular Professor Parry Jones of the King Fahd University of Petroleum and Minerals, Dhahran, Saudi Arabia, Dr R.S. Sayles of the Imperial College of Science, Medicine and Technology, University of London, D.R. White and M. Clarkson of the Measurement Standards Laboratory of New Zealand, and A. Maluyo and N. Liyanage for reviewing several chapters of the book and making valuable suggestions.

The assistance provided by Chandimal Fernando by drawing a large number of illustrations, Ms Roshini Thoradeniya by editing the script, Sunil Amarawansa, Kishantha Galappatti, Tharana Thoradeniya and Ms Imitha de Silva are gratefully acknowledged.

I am also grateful to the Institute of Measurement and Control, UK, the National Physical Laboratory, UK, the National Institute of Advanced Industrial Science and Technology, Japan, the National Metrology Institute of Japan, the National Institute of Standards and Technology (NIST), USA, the Commonwealth Scientific and Industrial Research Organization of Australia, the National Research Council of Canada, The International Organization for Standardization, the International Electrotechnical Commission, the International Organization for Legal Metrology, the American National Standards Institute, Fluke Corp., USA, Mitutoyo Corp., Japan, Isothermal Technology Ltd. UK, Morehouse Instrument Co. USA and Hottinger Baldwin Measurements (HBM) of Germany for providing materials and granting permission to reproduce them in the book.

Figures 3.3, 3.4, 3.5 and Table 3.1 (Critical characteristics of gauge blocks taken from ISO 3650: 1998), Table 6.5 (Classification of force proving instruments from ISO 376: 1987) and Table 6.7 (Classes of testing machines from ISO 7500-1: 1999) have been reproduced with the permission of the International Organization for Standardization, ISO. These standards can be obtained from any ISO member body or directly from the Central Secretariat, ISO, Case Postal 56, 1211 Geneva 20, Switzerland. Copyright remains with ISO.

The author thanks the International Electrotechnical Commission (IEC) for permission to reproduce extracts from IEC publications IEC 60751 (1983-01) entitled 'Industrial platinum resistance thermometer sensors' and IEC 60584-2 (1982-01) entitled 'Thermocouples. Part 2: Tolerances'. All extracts are the copyright of IEC, Geneva, Switzerland. All rights reserved. Further information on the IEC, its publications and its activities is available from www.iec.ch. IEC has no responsibility for the placement and context in which the extracts and contents are reproduced in this publication; nor is IEC in any way responsible for any of the other content or accuracy of this publication.

1

Requirements of ISO 9000 standards for test and measuring equipment

1.1 Introduction

Certification to ISO 9000 standards has become a primary requirement for both manufacturing and service-oriented organizations. Calibration and control of test, measurement and inspection equipment is one of the more important requirements given in the standard. A company that wants to obtain ISO 9000 certification therefore has to look into this vital aspect of their operations.

Test or calibration laboratories wishing to obtain independent third party certification should be guided by the requirements of ISO/IEC 17025: 1999 (formerly ISO/IEC Guide 25). A brief outline of the requirements of ISO 9001 standards is given in this chapter.

1.2 Evolution of ISO 9000 standards

The origin of the ISO 9000 series of quality management standards can be traced to the United States (US) military standards. The US military specifications MIL-I-Q9858 and MIL-I-45208 for quality inspection are the first standards to have specified requirements for quality assurance systems in the supplier's organization. Subsequently these standards were published as Allied Quality Assurance Publications (AQAP) 1, 4 and 9.

In 1972, the United Kingdom established UK Defence Standards 05/21, 05/24 and 05/29 based on the AQAP documents 1, 4 and 9. The famous British Standard BS 5750: 1979, parts 1, 2 and 3, were based on the presently obsolete UK Defence Standards 05/21, 05/24 and 05/29.

In 1985 the International Organization for Standardization through its Technical Committee on Quality Management and Assurance (ISO/TC 176) undertook the preparation of a series of international standards for quality management and BS 5750, which had been used successfully by the British Standards Institution for quality system certification, became the natural choice for basing the new international standard. After much deliberation and arguments ISO 9001, ISO 9002 and ISO 9003 were published in 1987. These standards were then adopted by a significant number of national standards bodies, including the United Kingdom, and were published as their national standards. The ISO 9000 series was also published as a European Standard series EN 29000 by the European Committee on Standardization (CEN).

In 1994 a revision of the series was undertaken, and an updated and revised set of standards was published. In the mean time a large number of organizations obtained certification against ISO 9001 and ISO 9002 standards. The usefulness of the standards for quality assurance of products and services was beginning to be accepted worldwide, though there were some organizations that were not entirely convinced by the necessity of a documented quality system as required by the standards.

A further revision of the standards was undertaken during 1996 to 2000, and a revised and improved set of standards known as ISO 9000: 2000 has been published. In the new standard certification can be obtained only against the ISO 9001 standard. ISO 9002 and ISO 9003 standards have been withdrawn. ISO 9004 has been published as a complementary guidance document.

1.3 Requirements of ISO 9001: 2000

The requirements of the ISO 9001: 2000 standard in respect of test, inspection and measuring equipment are summarized below:

(a) The organization shall identify the measurements to be made and the measuring and monitoring devices required to assure conformity of product to specified requirements.
(b) Measuring and monitoring devices shall be used and controlled to ensure that measurement capability is consistent with the measurement requirements.
(c) Measurement and monitoring shall be calibrated and adjusted periodically or prior to use, against devices traceable to international or national standards; where no such standards exist the basis used for calibration shall be recorded.
(d) Where applicable measuring and monitoring devices shall:

  (i) be safeguarded from adjustments that would invalidate the calibration;
  (ii) be protected from damage and deterioration during handling, maintenance and storage;
  (iii) have the results of their calibration recorded; and
  (iv) have the validity of previous results reassessed if they are subsequently found to be out of calibration, and corrective action taken.

Some guidelines for achieving these requirements are given. The international standard ISO 10012 – Part 1 is also a useful source of information for quality assurance of measuring equipment.

1.3.1 Identification of measurement parameters

This is the most important requirement from the point of view of product or service quality. This clause requires that the organization identifies the parameters of the product(s) for which tests or measurements should be carried out and that it equips itself adequately to carry out these tests and measurements. For most products the identification of the test parameters is relatively easy, as they are given in the product specification, national/international standard or specified by the customer.

However, equipping the organization to carry out the tests or measurements to the required level of accuracy is not straightforward as it may cost a relatively large sum of money to acquire the test equipment and for training of monitoring staff. To minimize costs it is necessary to obtain specialist advice as to the type of equipment that should be acquired. Training of the staff in the operation of the equipment and analysis of test data is also very important.

1.3.2 Measurement capability

The capability of the measuring instrument and procedure should not be less than the measurement requirement. Measurement requirements and capabilities are defined in terms of accuracy and uncertainty (see Chapter 2 for explanations of the term *accuracy* and *uncertainty*) of the measurement process, e.g. if a thickness measurement to an accuracy of ±0.1 mm is required, the instrument and the measurement procedure used for the purpose must be able to attain the same or slightly higher level of accuracy. In this instance the cheapest method would be to use a calibrated micrometer (with known corrections) with not more than ±0.02 mm calibration uncertainty.

1.3.3 Calibration of measurement and test equipment

Adjustment of a measuring instrument is an integral part of calibration. However, not all measuring instruments or artefacts are adjustable. Most length measuring instruments such as rulers, tapes and calipers are not adjustable. Also precision weights are not adjustable. For instruments that are non-adjustable, corrections are determined when they are calibrated. Thus when measurements are made using a non-adjustable instrument the correct procedure is to use the corrections given in the calibration certificate. The correction could be neglected, if it is smaller than the required accuracy by at least one order of magnitude. For example, if in the previous example the corrections of the micrometer are of the order of ±0.01 mm or less, then these corrections could be neglected as the required accuracy is only ±0.1 mm.

Traceability to international standards (see Chapter 2 for a discussion of *traceability*) is achieved by careful selection of the calibrating agency, making sure that their standards maintain traceability to international measurement standards. This is where laboratory accreditation comes into the picture as accreditation against ISO/IEC 17025 cannot be obtained without having established traceability to international measurement standards. Calibration laboratories to be used for ISO 9001 purposes should therefore have accreditation in terms of ISO/IEC 17025 standard.

National accreditation systems that accredit test and calibration laboratories are operated in many countries. The oldest accreditation body is found in Australia and is known as the National Accreditation and Testing Authority (NATA). The United Kingdom equivalent body is the United Kingdom Accreditation Service (UKAS) and in the United States there are at least two accrediting bodies, the American Association for Laboratory Accreditation (A2LA) and the National Voluntary Laboratory Accreditation Program (NVLAP).

Although not explicitly stated in ISO 9001, it is generally advisable to obtain calibration services from a laboratory accredited by the national accreditation service of the country, if such a system is available. Also international accreditation is available from bodies such as NATA and UKAS but these would be expensive for many organizations.

1.3.4 Recalibration interval

How is the period of recalibration to be determined? This depends on a number of factors, the most important of which are: the accuracy level, type of instrument and the frequency and conditions of use of the instrument (factory or laboratory). The organization should determine and document the recalibration intervals by analysing the past calibration records of the instrument and the drift observed. The manufacturer's recommendation in this regard is a useful guideline to follow. In the case of electrical measuring instruments the accuracies of instruments are usually given for a specific time period, e.g. 90 days, one year, etc. This means that the specified accuracy may be exceeded after the indicated time period. In such an event a recalibration may be required. Some statistical techniques have been developed to estimate recalibration intervals. The references to these are given in the Bibliography.

1.3.5 Sealing of adjusting mechanisms

Very often the calibration of an instrument is lost because someone had inadvertently adjusted the instrument. This often happens in the course of a minor repair, particularly in electrical measuring instruments where potentiometers are made available for adjustment purposes. It is a good practice to seal adjusting screws with security stickers or other appropriate means so that the instrument cannot be adjusted inadvertently. Generally all good calibration laboratories carry out such practices.

1.3.6 Handling and storage of test and measurement equipment

It is very important to handle test and measurement equipment with due care as their functioning and accuracy can deteriorate rapidly due to rough handling and inappropriate storage conditions, e.g. gauge blocks, micrometers, calipers and other length measuring instruments should be cleaned and returned to their boxes after use as ingress of dust can rapidly wear off their mating surfaces. Similarly precision weights should never be handled with bare hands. Cast iron weights should be cleaned and stored in a dust-free environment as dust particles act as nucleation centres for corrosion to take place thereby changing the value of the weight. Electrical measuring instruments are very susceptible to deterioration due to high temperature and humidity. In many countries with high humidity levels corrosion is a major problem. Due care such as applying a thin layer of oil or other protective material should be considered for long-term storage.

1.3.7 Documentation of calibration results

All calibrations carried out internally or by external laboratories should be documented in the form of a report. Essential details to be recorded are: date of calibration, item calibrated, reference standard used and its traceability, environmental conditions (temperature, humidity, etc.), a brief description of the calibration procedure or reference to the calibration procedure, details of results and uncertainties. A typical format of a calibration report is given in Appendix 1.

1.3.8 Discovery of out-of-calibration status

This very important aspect is often neglected by many organizations. If critical test equipment is found to be out of calibration, then the test results obtained using this piece of equipment for a considerable period of time prior to the discovery may have been inaccurate. It is necessary to launch an investigation to find out how this condition affected the product and the customer's requirements, and to take corrective action.

Bibliography

International standards

1. ISO 9001: 2000 Quality management systems – Requirements. International Organization for Standardization (ISO).
2. ISO 9004: 2000-1 Quality management systems – Guidelines for performance improvement. International Organization for Standardization (ISO).
3. ISO 19011 Guidelines for auditing management systems. International Organization for Standardization (ISO).
4. ISO 10005: 1995 Quality management – Guidelines for quality plans. International Organization for Standardization (ISO).
5. ISO 10006: 1997 Quality management – Guidelines to quality in project management. International Organization for Standardization (ISO).

6. ISO 10011-1: 1990 Guidelines for auditing quality systems – Part 1 Auditing. International Organization for Standardization (ISO).
7. ISO 10011-2: 1991 Guidelines for auditing quality systems – Part 2 Qualification criteria for quality systems auditors. International Organization for Standardization (ISO).
8. ISO 10011-3: 1991 Guidelines for auditing quality systems – Part 3 Management of audit programmes. International Organization for Standardization (ISO).
9. ISO 10012-1: 1992 Quality assurance requirements for measuring equipment – Part 1 Metrological confirmation system for measuring equipment. International Organization for Standardization (ISO).
10. ISO 10012-2: 1997 Quality assurance requirements for measuring equipment – Part 2 Guidelines for control of measurement processes. International Organization for Standardization (ISO).
11. ISO/IEC 17025: 1999 General requirements for the competence of testing and calibration laboratories. International Organization for Standardization (ISO).
12. International Document No. 10 – Guidelines for the determination of recalibration intervals of measuring equipment used in testing laboratories (1984) International Organization for Legal Metrology.
13. Establishment and Adjustment of Calibration Intervals – Recommended Practice RP-1 (1996) National Conference of Standards Laboratories.

2

Fundamental concepts of measurement

2.1 Introduction

Metrology or the science of measurement is a discipline that plays an important role in sustaining modern societies. It deals not only with the measurements that we make in day-to-day living, e.g. at the shop or the petrol station, but also in industry, science and technology. The technological advancement of the present-day world would not have been possible if not for the contribution made by metrologists all over the world to maintain accurate measurement systems.

The earliest metrological activity has been traced back to prehistoric times. For example, a beam balance dated to 5000 BC has been found in a tomb in Nagada in Egypt. It is well known that Sumerians and Babylonians had well-developed systems of numbers. The very high level of astronomy and advanced status of time measurement in these early Mesopotamian cultures contributed much to the development of science in later periods in the rest of the world. The colossal stupas (large hemispherical domes) of Anuradhapura and Polonnaruwa and the great tanks and canals of the hydraulic civilization bear ample testimony to the advanced system of linear and volume measurement that existed in ancient Sri Lanka.

There is evidence that well-established measurement systems existed in the Indus Valley and Mohenjedaro civilizations. In fact the number system we use today, known as the 'Indo-Arabic' numbers with positional notation for the symbols 1–9 and the concept of zero, was introduced into western societies by an English monk who translated the books of the Arab writer Al-Khawanizmi into Latin in the twelfth century.

In the modern world metrology plays a vital role to protect the consumer and to ensure that manufactured products conform to prescribed dimensional and quality standards. In many countries the implementation of a metrological system is carried out under three distinct headings or services, namely scientific, industrial and legal metrology.

Industrial metrology is mainly concerned with the measurement of length, mass, volume, temperature, pressure, voltage, current and a host of other physical and chemical parameters needed for industrial production and process control. The maintenance of accurate dimensional and other physical parameters of manufactured products to ensure that they conform to prescribed quality standards is another important function carried out by industrial metrology services.

Industrial metrology thus plays a vital role in the economic and industrial development of a country. It is often said that the level of industrial development of a country can be judged by the status of its metrology.

2.2 Fundamental concepts

The most important fundamental concepts of measurement except the concepts of *uncertainty of measurement* are explained in this section. The concepts of uncertainty are discussed in Chapter 9.

2.2.1 Measurand and influence quantity

The specific quantity determined in a measurement process is known as the *measurand*. A complete statement of the *measurand* also requires specification of other quantities, for example temperature, pressure, humidity, etc., which may affect the value of the *measurand*. These quantities are known as influence quantities.

For example, in an experiment performed to determine the density of a sample of water at a specific temperature (say 20°C), the *measurand* is the 'density of water at 20°C'. In this instance the only influence quantity specified is the temperature, namely 20°C.

2.2.2 True value (of a quantity)

The *true value* of a quantity is defined as the value consistent with its definition. This implies that there are no measurement errors in the realization of the definition. For example, the density of a substance is defined as mass per unit volume. If mass and volume of the substance could be determined without making measurement errors, then the true value of the density can be obtained. Unfortunately in practice both these quantities cannot be determined without experimental error. Therefore the *true value* of a quantity cannot be determined experimentally.

2.2.3 Nominal value and conventional true value

The *nominal value* is the approximate or rounded-off value of a material measure or characteristic of a measuring instrument. For example, when we refer to a resistor as 100 ohms or to a weight as 1 kg, we are using their nominal values. Their exact values, known as *conventional true values*, may be 99.98 ohms and 1.0001 kg respectively. The *conventional   true value* is obtained by comparing the test item with a higher level measurement standard under defined conditions. If we take the example of the 1 kg weight, the conventional true value is the mass value of the weight as defined in the OIML (International Organization for Legal Metrology) international recommendation RI 33, i.e. the apparent mass value of the weight, determined using weights of density 8000 kg/m^3 in air of density 1.2 kg/m^3 at 20°C with a specified uncertainty figure. The conventional value of a weight is usually expressed in the form 1.001 g ± 0.001 g.

2.2.4 Error and relative error of measurement

The difference between the result of a measurement and its *true value* is known as the *error* of the measurement. Since a *true value* cannot be determined, the *error* as defined cannot be determined as well. A *conventional true value* is therefore used in practice to determine an *error.*

The *relative error* is obtained by dividing the *error* by the average of the measured value. When it is necessary to distinguish *error* from *relative error*, the former is sometimes called *absolute error* of measurement. As the error could be positive or negative another term, absolute *value of error,* is used to express the magnitude (or modulus) of the error.

As an example, suppose we want to determine the error of a digital multimeter at a nominal voltage level of 10 volts DC. The multimeter is connected to a DC voltage standard supplying a voltage of 10 volts DC and the reading is noted down. The procedure is repeated several times, say five times. The mean of the five readings is calculated, and is found to be 10.2 volts.

The error is then calculated as 10.2 – 10.0 = +0.2 volts. The relative error is obtained by dividing 0.2 volts by 10.2 volts, giving 0.02. The relative error as a percentage is obtained by multiplying the relative error (0.02) by 100, i.e. the relative error is 0.2 per cent of the reading.

In this example a conventional true value is used, namely the voltage of 10 volts DC supplied by the voltage standard, to determine the error of the instrument.

2.2.5 Random error

The *random error* of measurement arises from unpredictable variations of one or more influence quantities. The effects of such variations are known as *random effects*. For example, in determining the length of a bar or gauge block, the variation of temperature of the environment gives rise to an error in the measured value. This error is due to a random effect, namely the unpredictable variation of the environmental temperature. It is not possible to compensate for random errors. However, the uncertainties arising from random effects can be quantified by repeating the experiment a number of times.

2.2.6 Systematic error

An error that occurs due to a more or less constant effect is a *systematic error*. If the zero of a measuring instrument has been shifted by a constant amount this would give rise to a systematic error. In measuring the voltage across a resistance using a voltmeter the finite impedance of the voltmeter often causes a systematic error. A correction can be computed if the impedance of the voltmeter and the value of the resistance are known.

Often, measuring instruments and systems are adjusted or calibrated using measurement standards and reference materials to eliminate systematic effects. However, the uncertainties associated with the standard or the reference material are incorporated in the uncertainty of the calibration.

2.2.7 Accuracy and precision

The terms *accuracy and precision* are often misunderstood or confused. The accuracy of a measurement is the degree of its closeness to the *true value*. The precision of a measurement is the degree of scatter of the measurement result, when the measurement is repeated a number of times under specified conditions.

In Fig. 2.1 the results obtained from a measurement experiment using a measuring instrument are plotted as a frequency distribution. The vertical

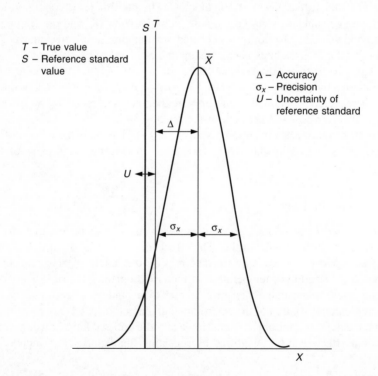

Figure 2.1 Accuracy, precision and true value

axis represents the frequency of the measurement result and the horizontal axis represents the values of the results (X). The central vertical line represents the mean value of all the measurement results. The vertical line marked T represents the *true value* of the measurand. The difference between the mean value and the T line is the *accuracy* of the measurement. The standard deviation (marked σ_x) of all the measurement results about the mean value is a quantitative measure for the precision of the measurement.

Unfortunately the accuracy defined in this manner cannot be determined, as the true value (T) of a measurement cannot be obtained due to errors prevalent in the measurement process. The only way to obtain an estimate of accuracy is to use a higher level measurement standard in place of the measuring instrument to perform the measurement and use the resulting mean value as the *true value*. This is what is usually done in practice. The line (S) represents the mean value obtained using a higher level measurement standard.

Thus accuracy figures quoted by instrument manufacturers in their technical literature is the difference between the measurement result displayed by the instrument and the value obtained when a higher level measurement standard is used to perform the measurement. In the case of simple instruments the accuracy indicated is usually the calibration accuracy, e.g in the calibration of a micrometer a series of gauge blocks is used. If the values displayed by the micrometer over its usable range falls within ±0.01 mm of the values assigned to the gauge blocks, then the accuracy of the micrometer is reported as ±0.01 mm.

It can be seen that the definition of *error* given previously (Section 2.2.4) is very similar to the definition of *accuracy*. In fact *error* and *accuracy* are interchangeable terms. Some prefer to use the term *error* and others prefer *accuracy*. Generally instrument manufacturers prefer the term *accuracy*, as they do not wish to highlight the fact that their instruments have errors.

Relative accuracy and per cent relative accuracy are also concepts in use. The definitions of these are similar to those of *relative error* and *per cent relative error,* i.e. *relative accuracy* is obtained by dividing *accuracy* by the average measured result and *per cent relative accuracy* is computed by multiplying relative accuracy by 100.

2.2.8 Calibration

Calibration is the process of comparing the indication of an instrument or the value of a material measure (e.g. value of a weight or graduations of a length measuring ruler) against values indicated by a measurement standard under specified conditions. In the process of calibration of an instrument or material measure the test item is either adjusted or correction factors are determined.

Not all instruments or material measures are adjustable. In case the instrument cannot be adjusted, it is possible to determine correction factors, although this method is not always satisfactory due to a number of reasons, the primary one being the non-linearity of response of most instruments.

For example, in the calibration of a mercury-in-glass thermometer between

0°C and 100°C, say the calibration was carried out at six test temperatures, 0°C, 20°C, 40°C, 60°C, 80°C and 100°C. Corrections are determined for each test temperature by taking the difference of readings between the test thermometer and the reference thermometer used for the calibration. These corrections are valid only at the temperatures of calibration. The corrections at intermediate temperatures cannot be determined by interpolation, e.g. the correction for 30°C cannot be determined by interpolating the corrections corresponding to 20°C and 40°C.

In the case of material measures, for example a test weight, either determination of the conventional mass value or adjustment of the mass value (in adjustable masses only) by addition or removal of material is performed. However, in the case of many other material measures such as metre rulers, gauge blocks, or standard resistors adjustment is not possible. In such cases the conventional value of the item is determined.

Some instruments used for measurement of electrical parameters are adjustable, e.g. multimeters, oscilloscopes, function generators.

2.2.9 Hierarchy of measurement standards

Measurement standards are categorized into different levels, namely primary, secondary and working standards forming a *hierarchy*. Primary standards have the highest metrological quality and their values are not referenced to other standards of the same quantity. For example, the International Prototype kilogram maintained at the International Bureau of Weights and Measures (BIPM) is the primary standard for mass measurement. This is the highest level standard for mass measurement and is not referenced to any further standard.

A *secondary standard* is a standard whose value is assigned by comparison with a primary standard of the same quantity. The national standard kilograms maintained by many countries are secondary standards as the value of these kilograms is determined by comparison to the primary standard kilogram maintained at the International Bureau of Weights and Measures (BIPM).

A standard, which is used routinely to calibrate or check measuring instruments or material measures, is known as a *working standard*. A working standard is periodically compared against a secondary standard of the same quantity. For example, the weights used for calibration of balances and other weights are working standards.

The terms *national primary standard*, *secondary standard* and *tertiary standard* are used to describe the *hierarchy* of national measurement standards maintained in a given country. Here the term *primary standard* is used in the sense that it is the highest-level standard maintained in a given country for a particular quantity. This standard may or may not be a primary standard in terms of the *metrological hierarchy* described in the previous paragraph, e.g. many countries maintain an iodine stabilized helium neon laser system for realization of the national metre. This is a case of a metrological primary standard being used as a national primary standard. On the other hand as pointed out earlier, the kilogram maintained by most countries as a national primary standard of mass is only a secondary standard in the metrological hierarchy of standards.

Usually the national hierarchy scheme is incorporated in the metrology law of the country.

A measurement standard recognized by international agreement to serve internationally as the basis for assigning values to other standards of the quantity concerned is known as an *international standard.*

The primary reason for establishing a hierarchy scheme is to minimize the use and handling of the higher level standards and thus to preserve their values. Thus the primary, secondary and working standards are graded in uncertainty, the primary standards having the best uncertainty and the working standards the worst uncertainty. Figure 2.2 depicts the two hierarchies of measurement standards.

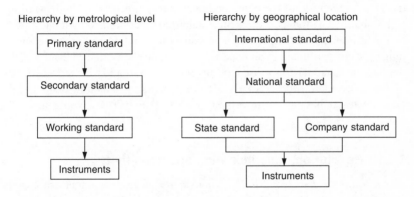

Figure 2.2 Hierarchies of measurement standards

2.2.10 Traceability

The concept of traceability is closely related to the hierarchy of standards. For a particular measurement standard, or measuring instrument, *traceability* means that its value has been determined by an *unbroken chain of comparisons* with a series of higher level standards with *stated uncertainties.* The higher level standards may be national standards maintained in a given country or international standards maintained by the International Bureau of Weights and Measures (BIPM) or any other laboratory.

Recently this fundamental definition has been modified by the addition of a time requirement for the comparisons. It is true that if the comparisons are widely separated in time, traceability may be lost. For example, a load cell fitted in a tensile testing machine may lose traceability after about one year from its last comparison. Thus the traceability of a test or measuring instrument depends largely on the type of instrument, the time interval from the last comparison and to some extent on the uncertainty of the instrument. Due to these reasons laboratory accreditation bodies such as the United Kingdom Accreditation Service (UKAS) and the National Association of Testing Authorities (NATA) of Australia have formulated specific guidelines for traceability of measurement standards and test and measuring equipment used in laboratories seeking accreditation.

2.2.11 Test uncertainty ratio (TUR)

Calibration of test and measurement equipment is always done against a higher level measurement standard, usually a working standard. The ratio of the *uncertainty* (see Chapter 9) of the test item to that of the measurement standard used in the calibration is known as the test uncertainty ratio. In most calibrations a TUR of at least 1:5 is used, though in some circumstances, especially when the test item has a relatively small uncertainty, a lesser TUR (1:2 or sometimes 1:1) has to be used. Nowadays it is more usual to determine the TUR as the ratio of the combined uncertainty (uncertainty budget) of the result obtained by the measurement standard to that obtained by the test item.

Let us look at an example. Say a pressure gauge of 0 to 1500 kPa (absolute) range is to be calibrated to an uncertainty of ±100 Pa. With a TUR of 1:5, the measurement standard to be used for this calibration should have an uncertainty of not more than 100/5 Pa i.e. ±20 Pa. A working standard dead weight pressure tester having an uncertainty of less than ±20 Pa would meet the criterion.

2.2.12 Resolution, discrimination and sensitivity

The *resolution, discrimination* and *sensitivity* of an instrument are closely related concepts. The resolution of a measuring instrument is the smallest difference between two indications of its display. For analogue instruments this is the smallest recognizable division on the display. For example, if the smallest graduation on a thermometer corresponds to 0.1°C, the resolution of the thermometer is 0.1°C. For a digital displaying device, this is the change in the indication when the least significant digit changes by one step, e.g. the resolution of a weighing balance indicating to two decimal places in grams is 0.01 g.

Discrimination on the other hand is the ability of an instrument to respond to small changes of the stimulus. It is defined as the largest change in a stimulus that produces no detectable change in the response of the measuring instrument. For example, if a mercury-in-glass thermometer is used to measure the temperature of an oil bath whose temperature is rising gradually, the smallest temperature change able to be registered by the thermometer will be its discrimination. This will not necessarily be equal to the resolution of the instrument. Generally in a good quality instrument the discrimination should be smaller than its resolution.

Sensitivity of an instrument is the numerical quantity that represents the ratio of the change in response to that of the change in the stimulus. This usually applies to instruments that do not have an output or response in the same units as that of the input or stimulus. A common example in a metrology laboratory is the equal arm balance. The input (stimulus) to the balance is the difference of mass between the two pans. The output is the angle of inclination of the balance beam at rest. Thus to relate the mass difference corresponding to a change in the angle of inclination of the balance beam, we need to determine the sensitivity of the balance. In the case of beam balances this is known as the sensitivity reciprocal.

2.2.13 Tolerance

Tolerance is the maximum allowable deviation of the value of a material measure, or the indication of a measuring instrument. In most cases tolerances are specified by national regulations or standard specifications. For example, the OIML International Recommendation RI 111 gives tolerances for weights of different classes used for metrological purposes (see Table 4.1).

2.2.14 Repeatability of measurements

The repeatability of a measuring instrument or measurement operation is defined as the closeness of the agreement between the results of successive measurements carried out under the same conditions of measurement within a relatively short interval of time. The repeatability conditions include the measurement procedure, the observer, the environmental conditions and location. Repeatability is usually expressed quantitatively as a standard deviation of the measurement result.

A familiar example is the repeatability of a weighing balance, which is determined by weighing a mass a number of times under similar conditions within a short interval of time. The standard deviation of the balance indications is expressed as the repeatability of the balance. A detailed procedure is given in Chapter 4.

2.2.15 Reproducibility of measurements

The reproducibility of a measurement process is the closeness of the agreement between the results of a measurement carried out under changed conditions of measurement. The changed conditions may include the principle of measurement, the method of measurement, the observer, the measuring instrument, the reference standards used, location where the measurement is performed, etc.

Reproducibility is rarely computed in metrology, though this concept is widely used and very useful in chemical and physical testing. Usually repeatability and reproducibility of a test procedure are determined by conducting a statistically designed experiment between two laboratories (or two sets of conditions) and by performing a variance analysis of the test results. The variance (square of the standard deviation) attributable to variation within a laboratory (or set of conditions) is expressed as repeatability and that between the laboratories is expressed as reproducibility. These experiments are usually known as R&R (repeatability and reproducibility) studies.

Bibliography

International standards

1. International vocabulary of basic and general terms in metrology (1993) International Organization for Standardization.
2. International Recommendation RI 33-1979. Conventional value of the result of weighing in air. International Organization of Legal Metrology.
3. International Recommendation RI 111-1994. Weights of classes E1, E2, F1, F2, M1, M2, M3. International Organization of Legal Metrology.

Introductory reading

1. De Vries, S. (1995) Make traceable calibration understandable in the industrial world. Proceedings of the Workshop on 'The Impact of Metrology on Global Trade'. National Conference of Standards Laboratories.
2. Sommer, K., Chappell, S.E. and Kochsiek, M. (2001) Calibration and verification, two procedures having comparable objectives, *Bulletin of the International Organization of Legal* Metrology, 17, 1.

Advanced reading

1. Ehrlich, C.D. and Rasberry, S.D. (1998) Metrological timelines in traceability, *Journal of Research of the National Institute of Standards and Technology,* 103, 93.

3

Linear and angular measurements

3.1 Introduction

Dimensional metrology is vitally important for the production of engineering components. Interchangeability of parts depends on the manufacture of components to predetermined tolerances. Linear and angular measurements are therefore of critical importance in the fields of production and mechanical engineering, and are also important in other areas such as architecture, civil engineering construction and surveying.

3.2 Length measurement

3.2.1 SI and other units

The SI base unit for length measurement is the metre defined as:

The length of the path travelled by light in vacuum during a time interval of 1/299 792 458 of a second.

This definition was adopted by the Conference Generale de Poids et Mesures (CGPM) in 1983 and fixes the velocity of light at 299 792 458 m/s.

A number of multiples and submultiples of the metre are in common use:

nanometre (nm) $= 10^{-9}$ metre
micrometre (μm) $= 10^{-6}$ metre
millimetre (mm) $= 10^{-3}$ metre
kilometre (km) $= 10^{3}$ metre

The centimetre (cm), widely used in the centimetre-gram-second system of units (CGS system), is not a recommended submultiple of the SI though this unit is still used in some situations.

The most widely used non-metric units are the inch, foot, yard and the mile. Presently, these units are also defined in terms of SI units as given below:

1 inch (in) = 25.4 millimetre exactly
1 foot (ft)   = 0.3048 metre
1 yard (yd) = 0.9144 metre
1 mile   = 1.609 34 kilometre

3.2.2 Primary standard

The metre has had a number of definitions since its original definition as one-ten millionth part of the meridian quadrant. When the metre convention was signed in 1875 it was defined as the distance between two marks made on a platinum iridium bar maintained at the BIPM. This standard is now only of historical importance.

The modern primary standard for length is the iodine stabilized helium neon laser. The frequency of this laser (f) is related to the wavelength of light (λ) through the relation:

$$c = f \lambda \tag{3.1}$$

where c = 299 792 458 m/s is the value assigned to the velocity of light by the definition of the metre. Length units are realized by incorporating the laser in an interferometer.

The frequency of the laser can be realized to an overall uncertainty of 1 part in 10^9. The frequency of the primary laser is disseminated to stabilized laser interferometer systems used for measurement of length and displacement.

3.2.3 Secondary and working standards

Helium neon lasers available commercially are widely used as the secondary standard for length. Usually these are used in combination with a measuring machine, gauge block comparator or coordinate measuring machine. In addition a range of other artefacts such as gauge blocks, length bars, line standards and tapes are used as working standards.

3.2.3.1 Measuring machine

A schematic diagram of a measuring machine is shown in Fig. 3.1. A measuring machine is a linear measuring instrument having a solid base and machined sliding surfaces. In the version known as the universal measuring machine linear measurements can be made in two axes perpendicular to each other.

Fixed and movable carriages are mounted on the sliding surfaces. In the older versions a line standard was incorporated in the machine. In modern versions a helium neon laser is mounted on the base as shown in Figure 3.1. A retro-reflector is mounted on the movable carriage. The laser beam is aligned so that it is parallel to the longitudinal axis of the machine. Interference takes place between the direct beam emitted by the laser and the beam reflected by the retro-reflector. By counting interference fringes, the distance between

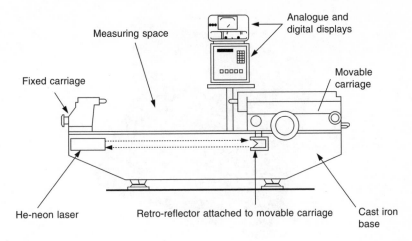

Figure 3.1 Measuring machine

the reference plane of the laser cavity and the retro-reflector is computed accurately.

Usually the fringes are counted electronically and converted to length units. An accurate value of the refractive index of air is required for this purpose. The temperature, pressure and relative humidity of ambient air are measured to calculate the refractive index. The frequency of the laser is obtained by comparison with a primary standard helium neon laser.

Measuring machines having a resolution of 0.1 μm and relative uncertainties of the order of 0.25 ppm are commercially available.

3.2.3.2 Gauge blocks

Gauge blocks were originally produced by C.E. Johansson of Sweden and are known as end standards as the reference surfaces of the block are its ends. Gauge blocks are available having rectangular or square sections. They are also available as short blocks (up to 100 mm) and long blocks (up to 1000 mm). Usually gauge blocks are available as sets consisting of a number of blocks enabling the building up of any length between 0.5 mm and 200 mm in steps of 0.0005 mm. A set of gauge blocks ranging from 2 mm to 100 mm consisting of 112 blocks is shown in Fig. 3.2.

Wringing of gauge blocks

Two gauge blocks can be wrung together to form an extended length. Wringing is done by bringing the surfaces of two blocks into contact and pressing lightly until they adhere to each other. The separation is done by carefully sliding one block parallel to the plane of adhesion. When two blocks are wrung together a contact layer of unknown thickness is formed between the two surfaces. The thickness of this layer depends on the surface roughness of

Figure 3.2 A set of steel gauge blocks. (Source: Mitutoyo Corp., Japan)

the two mating surfaces but for most practical purposes the effect of the wringing layer is insignificant. Before wringing gauge blocks together, their faces should be wiped free from dust and examined for burrs.

Building up a size combination of gauge blocks

The gauges to be used for building up a size combination are determined by the following method. The micrometer (0.001 mm) block is taken first, followed by the hundredth, tenth and millimetre blocks.

Example
To build up 108.455 mm from a 112-block set:

1st block   1.005
2nd block   1.05
3rd block   1.40
4th block   5
5th block   100

Documentary standards

The following international standards specify the dimensional and other requirements of gauge blocks:

ISO 3650: 1999 – Gauge blocks
OIML RI 30: 1981 – End standards of length (gauge blocks)

Also a number of national standards are listed in the Bibliography.

Critical characteristics

A number of important parameters of gauge blocks are defined in the international standards mentioned above. A summary of these is given below.

Length of a gauge block

The length (l) of a gauge block is the perpendicular distance between any particular point of one measuring face and the plane surface of an auxiliary plate of the same material and surface texture on which the opposite measuring surface of the gauge block has been wrung. The length includes one wringing film. This definition is illustrated in Fig. 3.3.

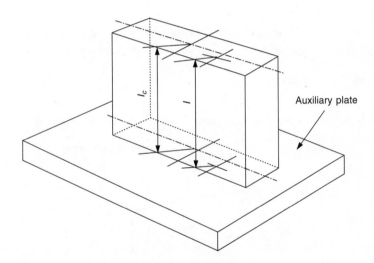

Auxiliary plate

Figure 3.3 Central length and length at any point of a gauge block. (Source: ISO 3650)

Central length

The central length (l_c) is a specific case of the length defined in the previous paragraph. The central length is the length taken at the centre point of the free measuring face.

Deviation from nominal length

The difference between the nominal length (l_n) and the length (l) at any point is the deviation from nominal length at any point, see Fig. 3.4.

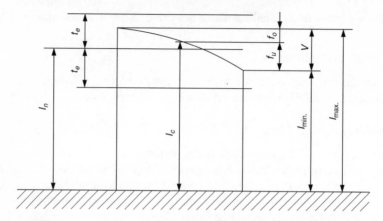

Figure 3.4 Geometrical properties of gauge blocks. (Source: ISO 3650)

Variation in length

The variation in length (V) is the difference between the maximum length ($l_{max.}$) and the minimum length ($l_{min.}$) of a given gauge block.

Deviation from flatness

The distance between two parallel planes enclosing all points of a measuring face is defined as the deviation from flatness. This is shown in Fig. 3.5.

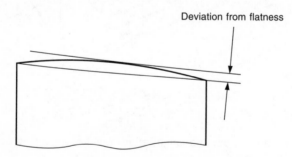

Figure 3.5 Deviation from flatness. (Source: ISO 3650)

Parallelism and perpendicularity of datum surfaces

The parallelism between the measuring faces and the perpendicularity of a measuring face with a side face are important parameters and are usually specified.

Surface finish

Surface finish of the measuring faces is important from the point of view of wringing quality so that the blocks wring readily. Fine scratches without burrs are acceptable provided they do not impair the wringing quality.

Dimensional stability

The stability of the length between the measuring surfaces of a gauge block over an extended period of time is an essential requirement. This is of the order of 0.02 μm per year for blocks of grade K and 0 and 0.05 μm for blocks of grade 1 and 2 respectively.

Material

Gauge blocks are made from hardened high grade alloy steel, zirconia ceramic or tungsten carbide. The hardness and long-term stability of the material are important properties. Gauge blocks made of steel should have a hardness of not less than 800 HV 0.5.

Blocks made from zirconia ceramic have excellent properties. The wringing characteristics are very good. They are also very tough and have very good wear resistance – up to three times that of tungsten carbide – low thermal conductivity, excellent resistance to corrosion and light weight. The coefficient of thermal expansion is close to that of steel ($9.5 \times 10^{-6}/°C$).

Tungsten carbide has also been used extensively as a material for the manufacture of gauge blocks. Gauge blocks made of tungsten carbide have very good wringing characteristics and good resistance to corrosion. The coefficient of thermal expansion ($4.23 \times 10^{-6}/°C$) is approximately half that of steel.

The lowest cost and most popular material for gauge blocks is hardened alloy steel. The major disadvantage of steel blocks is their susceptibility to corrosion.

Grades of gauge blocks

ISO 3650 specifies four grades of gauge blocks:

grade K, grade 0, grade 1 and grade 2

The grading is based on the deviation of length at any point from the nominal length as well as the tolerance for the variation in length. Grade K is specified as calibration grade for use as the reference standard for calibration of the other grades. Grade K gauge blocks will usually require calibration against a primary or secondary standard gauge block interferometer.

The ISO 3650 grading system is given in Table 3.1.

Care and use of gauge blocks

Recommendations made by the National Physical Laboratory[1] of the United Kingdom are given below.

---

[1]Crown copyright – Reproduced with the permission of the HMSO controller

Table 3.1 Limit deviations of length and tolerances from nominal length of gauge blocks. (Source: ISO 3650)

Nominal length l_n mm	Grade K		Grade 0		Grade 1		Grade 2	
	Limit deviation of length at any point from nominal length $\pm t_e$ μm	Tolerance for the variation in length t_v μm	Limit deviation of length at any point from nominal length $\pm t_e$ μm	Tolerance for the variation in length t_v μm	Limit deviation of length at any point from nominal length $\pm t_e$ μm	Tolerance for the variation in length t_v μm	Limit deviation of length at any point from nominal length $\pm t_e$ μm	Tolerance for the variation in length t_v μm
$0.5 \leq l_n \leq 10$	0.2	0.05	0.12	0.1	0.2	0.16	0.45	0.3
$10 < l_n \leq 25$	0.3	0.05	0.14	0.1	0.3	0.16	0.6	0.3
$25 < l_n \leq 50$	0.4	0.06	0.2	0.1	0.4	0.18	0.8	0.3
$50 < l_n \leq 75$	0.5	0.06	0.25	0.12	0.5	0.18	1	0.35
$75 < l_n \leq 100$	0.6	0.07	0.3	0.12	0.6	0.2	1.2	0.35
$100 < l_n \leq 150$	0.8	0.08	0.4	0.14	0.8	0.2	1.6	0.4
$150 < l_n \leq 200$	1	0.09	0.5	0.16	1	0.25	2	0.4
$200 < l_n \leq 250$	1.2	0.1	0.6	0.16	1.2	0.25	2.4	0.45
$250 < l_n \leq 300$	1.4	0.1	0.7	0.18	1.4	0.25	2.8	0.5
$300 < l_n \leq 400$	1.8	0.12	0.9	0.2	1.8	0.3	3.6	0.5
$400 < l_n \leq 500$	2.2	0.14	1.1	0.25	2.2	0.35	4.4	0.6
$500 < l_n \leq 600$	2.6	0.16	1.3	0.25	2.6	0.4	5	0.7
$600 < l_n \leq 700$	3	0.18	1.5	0.3	3	0.45	6	0.7
$700 < l_n \leq 800$	3.4	0.2	1.7	0.3	3.4	0.5	6.5	0.8
$800 < l_n \leq 900$	3.8	0.2	1.9	0.35	3.8	0.5	7.5	0.9
$900 < l_n \leq 1000$	4.2	0.25	2	0.4	4.2	0.6	8	1

(Reproduced from ISO 3650-1998 with permission of the International Organization for Standardization)

General care

The greatest care should be exercised in protecting gauge blocks and their case from dust, dirt and moisture. When not in actual use, the blocks should always be kept in their case and the case should be kept closed. The blocks should be used as far as possible in an atmosphere free from dust. In the case of steel blocks, care should be taken that the blocks do not become magnetized or they will attract ferrous dust.

Preparation before use

If the blocks are new or have been covered with a protective coating after being last used, most of this coating should be removed with an appropriate solvent (isopropyl or methyl alcohol). The measuring faces should finally be wiped with a clean chamois leather or soft linen cloth. This wiping should be carried out in every instance before a block is used, irrespective of whether it has been stored coated or merely returned temporarily to the case uncoated. It is, however, undesirable to aim at removing all traces of grease since a very slight film of grease is an aid to satisfactory wringing.

Care in use

Handling the lapped faces with bare hands should be avoided to reduce the risk of leaving fingerprints. Unnecessary handling of the blocks in use should be avoided as they take up the heat of the hand. If the blocks have been handled for some time they should be allowed to assume the prevailing temperature of the room before being used for test purposes. This is particularly important in the case of the larger sizes. When the highest accuracy is required, a test room with a controlled temperature of 20°C becomes necessary, but for ordinary purposes, provided the blocks and workpiece are of the same material, a sufficient degree of accuracy can be obtained if time is allowed to permit both to assume the prevailing temperature of the room.

Damaged gauges

Damage to the measuring faces is most likely to occur on the edges. Slight burrs may be removed with care by drawing an Arkansas type stone lightly across the damaged edge in a direction away from the measuring face of the standard. Any measuring face so treated should be thoroughly cleaned before wringing. A standard with a damaged measuring face should preferably be returned to the manufacturer for the surface to be restored.

Care after use

Immediately after use each block should be wiped clean and replaced in its proper compartment in the case. It is particularly important to remove any finger marks from the measuring faces. If the blocks are used infrequently they should be coated with a suitable corrosion preventive before being put away. The preparation should be applied to the measuring faces with a clean piece of soft linen. A brush should not be used as this may aerate the preparation and moisture in the air bubbles so formed can cause rusting of the faces.

3.2.4 Length measuring instruments

A brief description of length measuring instruments suitable for different purposes is given in this section.

3.2.4.1 Surface plate

The surface plate is an essential item for dimensional measurements both in the laboratory and in the factory. Usually all linear measurements are taken from a reference plane. The surface of a surface plate is used for this purpose.

Traditionally the flat surface plate was made from cast iron either as a freestanding table or suitable for mounting on a bench. Both had their upper surface hand scraped to a very high degree of flatness. The grade of surface plate is determined by the degree of flatness, which is defined as the distance between two parallel planes containing all the points of the surface. In recent years natural rock materials have become increasingly popular and granite plates have almost entirely replaced the cast iron plate.

Granite is twice as hard as cast iron and changes in temperature have only a minimal effect on its surface contour. The fine grain structure of well-seasoned granite ensures a surface largely free from burrs and protrusions and a high degree of flatness over a relatively long period of time.

Granite surface plates are available in sizes ranging from 300 mm × 300 mm × 100 mm to 2000 mm × 1500 mm × 300 mm. The best grade plates have a flatness ranging from 5 μm to 15 μm.

3.2.4.2 Outside micrometer

An outside micrometer (Fig. 3.6) is used to measure the thickness or diameter of hard materials. Outside micrometers are available with a measuring range of up to 2 m and 0.01 mm resolution. However, the range of the measuring head itself rarely exceeds 25 mm, and if the micrometer is required to measure large dimensions, it must be used as a comparator set either to zero position or preferably to a dimension near the workpiece size, using an end bar. Large size micrometers are usually available as a set, consisting of the frame, micrometer heads and setting bars.

A wide variety of these micrometers is available for different applications. Common types are with dial indicator, with LCD digital indicator, snap micrometer, dial snap micrometer, screw thread micrometer, tube micrometer, point micrometer and sheet metal micrometer.

Figure 3.6 Outside micrometer of 1 metre range. (Source: Mitutoyo Corp., Japan)

3.2.4.3 Inside micrometer

Micrometers used for internal measurements (internal diameters of cylinders and similar objects) are known as inside micrometers. Instruments with jaws range up to about 300 mm with a micrometer head of range 25 mm. Those exceeding a range of 300 mm are in the form of a cylinder with the measuring head at one end. Both types are available with a resolution of 0.01 mm. Inside micrometers having a measuring range of up to 5000 mm with micrometer head range of 50 mm are available. Cylindrical type inside micrometers are invariably supplied with sets of extensions that permit one head to cover a wide measuring range.

3.2.4.4 External and internal vernier calipers

Vernier calipers are used to measure external and internal diameters of objects and internal dimensions of cylinders and grooves. A range of sizes as well as a variety of types from the simple stainless steel type to those fitted with digital liquid crystal displays, battery or solar powered are available. Also, some types employ carbon fibre reinforced plastics in the beam and jaws to make them light yet strong. Most types are available up to a measuring range of 1000 mm with 0.01 mm resolution.

3.2.4.5 Dial gauge

A frequently used instrument for the measurement of small deviations relative to a datum surface is the dial gauge. A dial gauge consists of a spindle that moves inside a cylindrical tube. The linear motion of the spindle is translated into rotation of a shaft by the use of a rack and pinion mechanism. A pointer attached to the end of the shaft is made to move over a graduated circular dial.

In a different design, known as the lever type, the spindle is replaced with a stylus. The up and down motion of the stylus is translated into the rotary motion of the pointer.

A wide variety of dial gauges of different ranges, usually up to about 100 mm, and best resolution of 0.002 mm are available.

3.2.4.6 Bore gauge

A bore gauge is an instrument used for the measurement of the internal diameter of cylinders. A bore gauge consists of a cylindrical tube from which three prongs extend symmetrically. A micrometer head is attached to the cylinder and connected to the prongs internally. The movement of the micrometer spindle makes the prongs extend or contract. To measure the internal diameter of a cylinder at a given plane, the bore gauge is inserted into the cylinder until the prongs are positioned at the required plane. The micrometer spindle is rotated until the three prongs are in contact with the surface. The reading of the instrument gives the mean diameter from the three contact points.

Bore gauges are available in a wide variety of sizes and can be read to 0.001 mm.

3.2.4.7 Depth gauge

The depth gauge is another popular variation of a linear measuring instrument with a vernier reading and specialized anvils. The depth gauge is used to measure the depth of holes and steps.

3.2.4.8 Height gauge

The height gauge could be described as a vernier caliper fixed to a firm base. In contrast to the vernier caliper the height gauge has only one moving anvil connected to a vernier scale. A column having a fixed scale is mounted so that the axis of the column is perpendicular to the reference plane of the base. The motion of the anvil up and down the fixed scale allows vertical distances to be measured accurately. Usually a height gauge is mounted on a surface plate for measurement of vertical distances.

3.2.4.9 Tapes

Steel and fabric tapes are used for measurement of lengths in excess of one metre. Usually steel tapes are available in lengths of 50 m, 100 m and up to about 500 m. Fabric tapes are available up to lengths of 100 m.

3.2.4.10 Laser measuring systems

A number of linear measuring systems using laser interferometry are available. In most systems a helium neon laser operating at 633 nm is used as the coherent source of light. These systems have very good resolution and accuracy as well as other features such as non-contact measurement capability. Also data acquisition and analysis are conveniently handled by either inbuilt processors and programs or external hardware and software. Laser interferometry systems have measuring ranges of up to 30 metres with resolution of a few micrometres.

Instruments based on the principle of simple interference of a direct and reference beam of light are excellent instruments for linear measurements. However, they require an accurate value for the refractive index of air to compute the length from the measured phase difference of the two interfering beams. Though this is relatively easy under laboratory conditions, in the field or in a factory the same accuracy cannot be obtained due to the variation of ambient temperature, humidity and pressure under field conditions.

A laser grating interferometer is less susceptible to variation of the refractive index of air and is an ideal instrument for use in industrial situations. Grating interferometers measure length using the interference of two light beams diffracted on a diffraction grating and subsequent evaluation of the phase difference. Linear encoders fitted to many linear measuring instruments is based on this principle. Linear encoders incorporating laser-grating interferometers have a maximum measuring range of about 1500 mm with an uncertainty of ±10 µm.

A laser scan micrometer using the principle of scanning laser interferometry is shown in Fig. 3.7. This type of instrument is capable of measuring workpieces that are brittle or soft and which may suffer dimensional change due to the measuring force. Also these instruments easily handle workpieces that are at a raised temperature and difficult for measurement by conventional instruments.

Laser scan micrometers are available with measuring ranges of 2 mm to 120 mm and resolutions of 0.01 µm to 0.1 µm. Repeatability of these instruments is of the order of ±2 µm.

Figure 3.7 Laser scan micrometer. (Source: Mitutoyo Corp., Japan)

3.2.5 Coordinate measuring machine (CMM)

A coordinate measuring machine (CMM) is a general-purpose, high-speed instrument used for measuring small to medium sized workpieces. These machines offer high measurement accuracy and excellent measuring efficiency. Most modern instruments are automated with inbuilt or external computers and are relatively simple to operate.

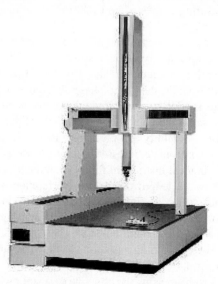

Figure 3.8 Coordinate measuring machine. (Source: Mitutoyo Corp., Japan)

The essential components of a coordinate measuring machine are a work table and carriages movable in three mutually perpendicular axes (X, Y and Z). The main carriage is movable along the longitudinal axis (X axis) of the work table. A secondary carriage (Y axis) is movable on the crossbar (bridge) of the X-axis carriage and a vertically mounted spindle constitutes the Z axis. A probe is attached to the end of the Z-axis spindle.

The motions of the three carriages in X (back and forth), Y (right and left) and Z (up and down) directions enable the probe to be placed at any point within the measuring volume of the instrument. Precise linear encoders, installed on each axis, measure the probe displacement. The positional data obtained at each measured point is output to a computer for two- or three-dimensional measurement including determination of point coordinates, dimensions and contours. These abilities are especially useful when measuring bulky workpieces that are hard to handle, and when making multi-planar measurements of complicated workpieces.

Coordinated measuring machines are best suited for measuring fabricated products such as:

(a) Moulds for pressing, die casting injection moulding, and precision castings.
(b) Moulded products before and after machining.
(c) Prototypes made on numerically controlled machine tools.
(d) Machined parts.

3.2.5.1 Advantages of coordinate measuring machines

Coordinate measuring machines have the following advantages over conventional measuring systems:

(a) The coordinates of any point on a workpiece are determined simply by placing the probe tip in contact with it.
(b) All workpiece faces, other than the bottom, may be measured provided the workpiece is correctly mounted.
(c) Highly precise measurements can be performed with minimal training. In addition, the time required for workpiece set-up and measurement is reduced.
(d) Datum (reference) points can be specified as required.
(e) A data processing unit instantly determines and prints dimensions, co-ordinates and contours of workpieces, and then compares the measurements with their design values and tolerance limits, resulting in a great reduction in time for measurement data analysis.

The accuracy of measurement of a coordinate measuring machine depends on the nominal dimensions of the workpiece. Typical accuracies and uncertainties of a commercially available instrument are given below:

Linear displacement accuracy: $3 + 3L/1000$ μm
Axial length measuring accuracy: $3 + 3L/1000$ μm
Volumetric length measuring accuracy: $3.5 + 3.5L/1000$ μm
Repeatability ± 6 μm

where L = nominal length in metres

3.3 Calibration of dimensional standards and measuring instruments

3.3.1 Measurement errors and their correction

3.3.1.1 Effect of temperature

Temperature of the environment is a critical influence factor for dimensional measurements. The recommended standard temperature is 20°C. However, even in a laboratory with temperature controlled at 20°C, the workpiece temperature may be slightly different. Therefore the temperature of the workpiece is measured using a calibrated thermometer and the dimensional values measured are corrected using the thermal expansion coefficient of the material.

The change in length (δL) of a test item due to a change in temperature is given by the following equation:

$$\delta L = L \cdot \alpha \cdot \delta t \qquad (3.2)$$

where:

L = original length of the test item
α = linear expansion coefficient of the material
δt = change in temperature

The linear thermal expansion coefficients of several materials are given in Table 3.2. When the measuring instrument and the test item being measured are of two different materials a differential expansion results. Corrections are required for these effects.

Table 3.2 Linear thermal expansion coefficients of materials

Material	Linear expansion co-efficient, /°C
Cast iron	$9.2-11.8 \times 10^{-6}$
Steel	11.5×10^{-6}
Chromium steel	$11-13 \times 10^{-6}$
Nickel chromium steel	$13-15 \times 10^{-6}$
Copper	18.5×10^{-6}
Bronze	17.5×10^{-6}
Gunmetal	18.0×10^{-6}
Aluminium	23.8×10^{-6}
Brass	18.5×10^{-6}
Nickel	13.0×10^{-6}
Iron	12.2×10^{-6}
Nickel steel	12.0×10^{-6}
Invar (36% nickel)	1.5×10^{-6}
Gold	14.2×10^{-6}
Glass (crown)	8.9×10^{-6}
Glass (Flint)	7.9×10^{-6}
Ceramics	3.0×10^{-6}

3.3.1.2 Deformation

Deformation is the second most important influence factor. Mainly three effects cause deformations:

(a) Due to force exerted on the test piece by the measuring instrument.
(b) The method of support used for the test item.
(c) The method of support of the measuring instrument.

Force exerted by the measuring instrument

The deformation (within the elastic limit) due to the force exerted by the measuring instrument is estimated from:

$$\Delta L = \frac{F \cdot L}{E \cdot A} \qquad (3.3)$$

where:
F = measuring force
L = length of test item
E = Young's modulus
A = cross-sectional area of the test item.

The method of support of the test item

When a test item is supported on a flat surface such as a surface plate the unevenness of the plate will affect the measurements to be performed. Due to these reasons and for precise measurements test items are generally supported on knife edges or cylinders. The knife edges or cylinders are placed symmetrically in relation to the length of the test item.

Airy points

At Airy points the two vertical edges of an end standard remain parallel to each other, Fig. 3.9(a).

The distance between the Airy points is given by:

$$d = \frac{L}{\sqrt{n^2 - 1}} \qquad (3.4)$$

where:
L = length of the test piece
d = distance between Airy points
n = number of supports.

For two point support, $d = 0.5774L$ and $a = 0.2113L$.

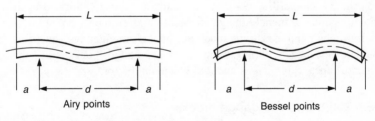

Figure 3.9 Airy and Bessel points of support

Bessel points

When a long test item is supported at Bessel points the contraction of the centre line of the item due to its own weight is a minimum, Fig. 3.9(b). This method of support is most suitable for calibration of long standard scales.

The distance between Bessel points is $0.5594L$ and $a = 0.2203L$.

Abbe error

Abbe error arises when the axis of the instrument or standard being used for the measurement does not coincide with the axis of the test item being measured. The deformation of the measuring arm of a caliper type micrometer through an angle θ gives rise to an error in length of $d\theta$.

e.g if $\theta = 1$ minute and $d = 30$ mm,

Abbe error $= 30 \times 1/3000 = 10$ μm

3.3.1.3 Parallax error

Parallax errors occur when viewing a pointer against a graduation mark as in the case of a dial indicator or viewing graduations that are in different planes as in the case of a micrometer sleeve and thimble. In these situations it is best to observe normal to the plane of the graduations as far as practicable.

3.3.2 Reference conditions

The reference temperature for dimensional measurements is 20°C. If a temperature different from 20°C is used a correction using the linear coefficient of expansion of the material is made (see Table 3.2). All test items such as gauge blocks, end bars, micrometers and vernier calipers should be kept in the temperature controlled room, for at least one hour prior to the commencement of calibration to attain temperature equalization.

3.3.3 Reference standard

The hierarchy of length measurement standards is given in Fig. 3.10. This diagram gives the next higher level standard that may be used for the calibration of a given standard or instrument. The iodine stabilized helium neon laser is the primary standard. Secondary standards are those calibrated against the primary standard, namely the gauge block interferometer, helium neon laser interferometer and measuring machine, and helium neon laser interferometer coupled tape bench. Working standards are gauge blocks, line standards, standard tapes, ring gauges and dial gauge calibrators.

3.3.4 Calibration of gauge blocks

The highest level gauge blocks, that is those of ISO 3650 grade K and OIML

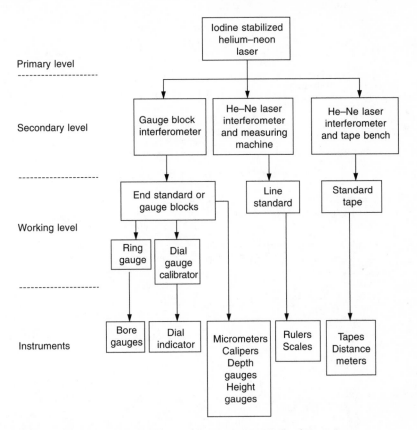

Figure 3.10 Hierarchy of length measurement standards

class AA are calibrated against a secondary standard, i.e. He–Ne laser gauge block interferometer. Working standard gauge blocks are calibrated by comparing them against a set of grade K gauge blocks in a gauge block comparator, Fig. 3.11.

The comparison is carried out as a one to one comparison or as a multi-measurement procedure in a least squared design. In a one to one comparison gauge blocks of the same nominal length are compared and the deviation of the block from its nominal value is determined using the value of the known block.

In a least squared design a number of blocks of a set are measured using the comparator and the deviations are determined by multi-linear regression. Details of these procedures are beyond the scope of this book but may be found in the references given in the Bibliography.

3.3.5 Calibration of micrometers

The calibration of a micrometer consists of testing the instrument for accuracy, flatness of measuring faces and parallelism of measuring faces.

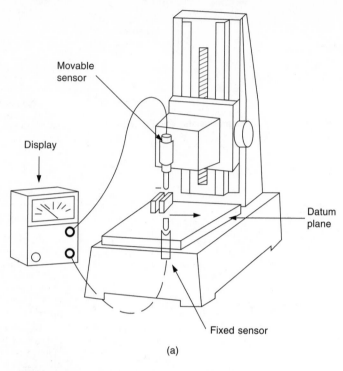

Movable
sensor

Display

Datum
plane

Fixed sensor

(a)

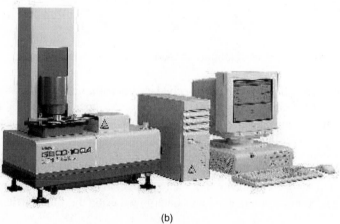

(b)

Figure 3.11 Gauge block comparators: (a) mechanical comparator; (b) optical comparator. (Source: Mitutoyo Corp., Japan)

3.3.5.1 Deviation of reading

The accuracy of a micrometer over its working range is determined by measuring a set of gauge blocks of appropriate class, usually OIML RI 30 class B or ISO 3650 grade 1 or 2, using the micrometer under calibration. The micrometer is firmly held in a rigid stand and the gauge block is introduced between the anvils. The movable anvil is rotated until it comes in contact with the datum surface of the gauge block. The micrometer reading is taken and compared with the actual length of the gauge block, obtained from its calibration certificate. The procedure is repeated at least three times at each major graduation and the mean value of the reading computed.

3.3.5.2 Flatness of measuring faces

The flatness of measuring faces is tested by using an optical flat, Fig. 3.12. An optical flat of diameter 45 mm or 60 mm is generally used. After cleaning the measuring face of the anvil thoroughly, one surface of the optical flat is brought into contact with the measuring face. Unless the faces are perfectly flat a number of coloured interference bands will be seen on their surfaces. The shape and the number of these bands indicate the degree of flatness of the face. A band represents a distance of 0.32 μm.

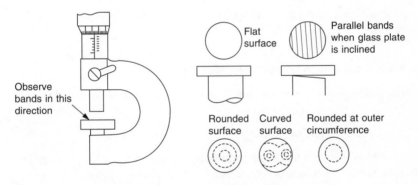

Figure 3.12 Testing of flatness of a micrometer measuring face

3.3.5.3 Parallelism of measuring faces

Parallelism of measuring faces is determined by either using optical parallels or gauge blocks. Parallelism should be tested at four angular positions of the anvil. For 0–25 mm micrometers, 12.00 mm, 12.12 mm, 12.25 mm and 12.37 mm parallels are used.

The optical parallel is first placed on the measuring face of the anvil and carefully moved until the bands visible on the face are reduced to a minimum. The measuring face of the spindle is then brought into contact with the optical parallel. The number of bands visible on both faces gives the parallelism of the measuring faces.

For micrometers of range 50 mm and larger it is more convenient to use gauge blocks. A gauge block is placed at five positions between two measuring faces. The maximum difference of the five readings is considered as the parallelism of the measuring faces. Testing is preferably done at two angular

Figure 3.13 A set of optical parallels. (Source Mitutoyo Corp., Japan)

positions of the anvil, by using two gauge blocks differing in length by 0.25 mm.

3.3.6 Calibration of vernier calipers

The calibration of a vernier caliper basically consists of the following tests.

3.3.6.1 Deviation of reading

The deviation of reading is determined by the use of gauge blocks or end bars or by the use of a measuring machine. The deviations of reading at not less than five positions, equally spaced within the measuring range of the main scale and the vernier scale, are determined.

3.3.6.2 Flatness of the measuring faces

The flatness of the measuring faces for both external and internal measurement is determined by using either a dial test indicator or an optical flat. When a dial test indicator is used, the test instrument is laid horizontally on a surface plate and the tip of the dial test indicator is traversed along the surface of the measuring face. The maximum deviation of the indicator is taken as a measure of flatness. When an optical plate is used, a procedure similar to testing of flatness of the measuring faces of a micrometer is followed.

3.3.6.3 Parallelism of the measuring faces

Parallelism of the measuring faces is determined by inserting gauge blocks at different points on the jaws or by using a measuring machine. Generally parallelism is determined at two measured lengths, mid range and close to full range.

3.3.6.4 Squareness of the fixed face

The squareness of the fixed face for external measurement with the guiding edge of the beam is determined by holding a gauge block of comparable length against the edge of the beam and fixed measuring face.

In addition the thickness of the main scale lines and vernier scale lines are checked by direct measurement using an optical microscope.

3.3.7 Calibration of dial gauge

3.3.7.1 Deviation of reading

The deviation of reading of a dial gauge is usually determined using a dial gauge calibrator. A simple design of a dial gauge calibrator is shown in Fig. 3.14. The instrument consists of a calibrated micrometer head attached to a fixed vertical column. The dial gauge under test is clamped and held rigidly opposite, and in line with the axis of the micrometer head.

A series of readings is taken at suitable intervals throughout the range of the gauge. If the gauge has only a limited range with two or three turns of its

Figure 3.14 Dial gauge calibrator (Source: Mitutoyo Corp., Japan)

pointer a tenth of the range is a suitable interval. In the case of gauges with longer scales a few intervals in each revolution is tested in order to keep the number of readings within reasonable bounds.

3.3.7.2 Repeatability of reading

The repeatability of a dial gauge could be determined in one of two ways. The dial gauge is firmly clamped in a suitable rigid fixture over a flat steel base and a cylinder is rolled under the contact point a number of times from various directions. The test is repeated at two or three points along the range of the gauge.

Alternatively the contact point is allowed to rest on a flat surface below and the plunger is lowered on to the surface both slowly and abruptly. In both cases the largest difference between the readings is noted as the repeatability.

3.3.7.3 Discrimination

Stickiness or backlash of a dial gauge is revealed by the test for discrimination. The dial gauge is mounted in a rigid fixture with the end of its plunger in contact with the surface of a slightly eccentric precision mandrel mounted between centres. A sensitive indicator (e.g. roundness measuring machine) is used beforehand to determine the amount of eccentricity in the mandrel.

3.3.8 Calibration of ring gauges

A ring gauge or setting ring is calibrated in a ring gauge calibrator. Two versions of ring gauge calibrators are available, the mechanical comparator type where the ring gauge under calibration is compared with a block of gauge blocks and the comparator of more recent origin where the comparison is against a laser holoscale or grating interferometer. An instrument incorporating a laser holoscale is shown in Fig. 3.15.

A number of diameters of the ring gauge are measured and the mean computed. Also the out of roundness or ovality of the ring gauge is estimated.

3.4 Measurement of angle

3.4.1 SI and other units

3.4.1.1 Plane angle

The radian (rad) is the SI unit for the measurement of plane angle. It is defined as:

Figure 3.15 Ring gauge calibrator (Source: Mitutoyo Corp., Japan)

the plane angle subtended at the centre of a circle by an arc equal to the radius of the circle.

The degree, second and minute are also in common use for the measurement of plane angle. The relationships among these units are given below:

1 radian (rad) = 180 degrees (°)
1 degree = 60 minutes (')
1 minute = 60 seconds (")

3.4.1.2 Solid angle

The steradian (sr) is the SI unit for measurement of solid angle. It is defined as:

that solid angle, having its vertex at the centre of a sphere, cuts off an area of the surface of the sphere equal to that of a square with sides of length equal to the radius of the sphere.

From the above definition, the solid angle subtended at the centre of a sphere by its entire surface is computed as 4π steradians.

The radian and steradian are special names given to the two non-dimensional derived units.

3.4.2 Angle standards

3.4.2.1 Plane angle

The radian can be realized in a fundamental way by measuring the radius and arc of a circle. The uncertainty of this determination would be governed by the uncertainties of measuring the two lengths. However, due to practical difficulties of measuring curved lengths this method is not attempted in practice and other methods of generating known plane angles are used.

The instruments used at the primary level for generation and measurement of angles are the sine bar, the indexing table and regular polygon.

3.4.2.2 Sine bar

A sine bar consists of a flat rectangular section to which two rollers are rigidly attached. The distance between the rollers is very important and determines the accuracy of the instrument. This distance is usually in multiples of 50 mm for ease of calculation.

The method of using a sine bar to generate an angle is shown in Fig. 3.16. The vertical dimension of the triangle is formed by gauge blocks of such a length that division by the length of the sine bar gives the sine of the required angle.

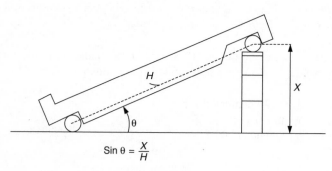

$$\mathrm{Sin}\,\theta = \frac{X}{H}$$

Figure 3.16 Use of a sine bar to generate an angle

This method is difficult to use as a direct means of measurement but is used for generation of accurate angles. A useful development of the sine bar is the sine table, which is formed by increasing the width of the sine bar. This provides a plane on which the workpiece can be mounted. A further development is the compound sine table, which is effectively two sine tables mounted at right angles on top of each other. A complex workpiece can be mounted on the upper sine table so that a chosen surface is made parallel to the base plane.

3.4.2.3 Indexing table

The most popular instrument for angle generation and measurement is the indexing table. The indexing table is a rotating table made by dividing a circle accurately. Both motor driven and hand driven types are available. In the most precise instruments angles can be generated and measured to the nearest second.

3.4.2.4 Precision polygon

Precision polygons are made from hardened and stabilized steel or glass. They have lapped or polished working surfaces normal to equal divisions of a circle. Polygons with 12 faces at 30° intervals or 72 faces at 5° intervals are normally available. Precision polygons are mainly used for calibrating indexing tables and dividing heads. In the calibration of an indexing table the polygon is mounted on the table with its axis coincident with the axis of the table and the table is rotated through the angle of the polygon. By focusing an autocollimator on two adjacent working faces of the polygon the deviation of the rotation of the table in relation to the angle of the polygon is determined.

3.4.2.5 Angle gauge

An angle gauge is a block of hardened steel with two lapped working surfaces at a precise angle to each other. They are available in sets and can be wrung together to form angles when used together with a precision square block.

A system of angle gauges invented by H. Tomlinson of the National Physical Laboratory of the United Kingdom has 12 angle gauges of nominal values:

3, 9, 27 seconds
1, 3, 9, 27 minutes
1, 3, 9, 27, 41 degrees

and a single square block. This set provides angles up to 360° with an interval of 1.5°. The angle gauges are approximately 75 mm long and 16 mm wide and can be wrung together to form either additive or subtractive combinations.

3.4.2.6 Autocollimator

An important instrument used for angle measurement is the autocollimator. A collimating lens is used to focus a parallel beam of light onto a reflecting surface kept normal to the incident beam of light and the reflected beam is focused onto a measuring eyepiece. If the reflector is inclined at a small angle $\delta\theta$ to the normal, the reflected beam is inclined at an angle of $2\delta\theta$ to the transmission path and is measured by a scale incorporated in the microscope eyepiece. An autocollimator in combination with a precision polygon is used to calibrate indexing tables.

Bibliography

International and National standards

1. ISO 3650: 1998. Length standards – Gauge blocks. International Organization for Standardization.
2. OIML RI 30: 1981. End standards of length. International Organization of Legal Metrology.
3. BS 4311: 1977. Gauge blocks. British Standards Institution.
4. BS 5317: 1976. Length bars. British Standards Institution.
5. BS 4064: 1994. Plain setting rings. British Standards Institution.

Introductory reading

1. Collett, C.V. and Hope, A.D. (1983) *Engineering Meaurements*, Longman Scientific and Technical.
2. Gayer, J.F.W. and Shotbolt, C.R. (1990) *Metrology for Engineers*. Cassell Publishers.

Advanced reading

1. National Bureau of Standards NBSIR 76-979 (1978) *Intercomparison Procedures for Gauge Blocks using Electromechanical Comparators*, US Department of Printing.

4

Mass measurements

4.1 Introduction

Mass measurement is practised widely both in industry and in trade and commerce. Mass is defined as the 'quantity of matter of an object'. However, the measurement of mass is mostly carried out by weighing, using a variety of balances.

4.2 Primary standard and SI units

The SI base unit for mass is the kilogram (kg), defined as the mass of the international prototype kilogram maintained at the Bureau des Poids et Mesure (BIPM) in Sevres, France. The kilogram remains the only artefact standard of the modern SI system of units. The primary standard of mass is also the international prototype kilogram, made from platinum iridium alloy.

4.3 Secondary and working standards

At the signing of the Treaty of the Metre, 48 copies of the kilogram were made and distributed to the 48 national laboratories of the member countries. These kilograms constitute the secondary standards of mass. Countries that joined the metre convention later were also given a copy of the kilogram.

In addition national laboratories maintain a set of weights known as tertiary standards. These standards are used for the calibration of weights used in industry and trade and commerce, except in those circumstances where uncertainty of the weights requiring calibration demands a higher level standard to be used.

4.4 Mass and weight

Mass and weight are often confused as synonymous, though they are two distinctly different quantities. Mass is defined as the amount of matter in an object. It is also defined using Newton's second law of motion, namely:

$$F = m \times a \tag{4.1}$$

Force = mass × acceleration

Each object possesses a property called *mass*, which appears in the equation as the constant of proportionality between a force (F) applied to the object of mass (m) and the resulting acceleration (a) of the object.

The *weight* of an object is the force experienced by it due to the earth's gravity. Since an object of mass (m) will accelerate through 'g', the acceleration due to gravity, we can write the above equation for motion under gravity:

$$W = m \times g \tag{4.2}$$

W is known as the weight of the object. Thus, the weight of an object would vary by a very small amount, as it moved about from place to place, on the surface of the earth, due to the variation of the value of 'g'.

Weight being a force should be measured in force units, in SI the unit is the newton (N). In the older metric systems of units, namely the centimetre-gram-second (CGS) and metre-kilogram-second (MKS) systems, the units 'gram-weight' and 'kilogram-weight' were used. However, a problem arose due to these units, defined as the force exerted by the earth's gravity on a mass of one gram or one kilogram respectively, being dependent on the value of 'g', the gravitational acceleration. To overcome these difficulties, a unit for measurement of force known as the 'kilogram-force' was defined. The kilogram-force is the force experienced by a kilogram due to a standard acceleration of $9.806\,55$ m/s^2.

In coherent SI units, these difficulties do not arise, as mass and force are measured using distinctly recognizable non-gravitational units, namely the kilogram and the newton. The newton is a derived unit of the SI, defined as the force required to accelerate a mass of one kilogram through 1 m/s^2.

4.4.1 True mass (vacuum mass)

True mass of an object is the mass as defined by equation (4.1). If the mass of an object is determined by weighing it in a vacuum, the mass obtained will be identical to that obtained using equation (4.1). Due to this reason 'true mass' is also known as 'vacuum mass'. However, weighing in a vacuum is rarely carried out even in the advanced metrology laboratories though the concept is very useful to explain mass measurement procedures.

4.4.2 Air buoyancy effects and apparent mass

If an object is weighed in a medium such as air (or a liquid), it experiences a force known as *up thrust*. The *up thrust* is numerically equal to the *weight* of the fluid displaced by the object. This is known as Archimedes' principle.

If the density of the fluid is d and the volume of the object is V, the up thrust U is given by:

$$U = d \times V \times g \qquad (4.3)$$

Thus, if an object of true mass M_t is weighed on a direct reading balance, the mass registered by the balance, M_x is given by:

$$M_x = M_t - d \times V \qquad (4.4)$$

The true mass is reduced by the term $(d \times V)$. $(d \times V)$ is known as the *buoyancy correction*. Mx is loosely referred to as the *apparent mass* of the object. A more complete definition of the *apparent mass* of an object is given below:

The apparent mass M_X^A of an object X is equal to the true mass M_R^T of just enough reference material to produce a balance reading equal to that produced by X if the measurements are done at temperature t_o in air of density d_o.

Apparent mass M_X^A is related to the True Mass M_X^T by the following equation:

$$M_X^A = M_X^T \frac{(1 - d_a/d_x)}{(1 - d_a/d_R)} \qquad (4.5)$$

where:
d_a = density of air in which the weighing is carried out
d_x = density of object X
d_R = density of reference material.

4.4.2.1 Reference materials

At present two different apparent mass bases are used internationally. The older one of these is called *normal brass* and was the logical choice when most laboratory weights were made of brass.

Normal brass, which is also known as 8.4 system, is defined by:

8.4 System
$d_R = 8400 \ kg/m^3$ at $0 \, ^\circ C$
$\alpha_R = 5.4 \times 10^{-5}/^\circ C$ = coefficient of cubical expansion
$d_a = 1.2 \ kg/m^3$ = air density at $20 \, ^\circ C$

8.0 System
The second apparent mass base is referenced to an arbitrary material of density 8000 kg/m^3 at 20°C and air density of 1.2 kg/m^3 at 20°C. On this base, the coefficient of cubical expansion is not needed as both densities are referenced to 20°C.

4.4.2.2 Conventional mass value

The conventional mass value of an object is a specific apparent mass, where the density of the reference material and the density of air are defined to be

8000 kg/m^3 and 1.2 kg/m^3 at 20°C respectively (OIML RI 33 – Conventional value of the result of weighing in air).

4.4.2.3 Relationship between true mass and conventional mass

If an object X of true mass, M_X^T is weighed using weights of conventional mass value M_x^8 in air of density d_a:

$$M_x^T = M_x^8 \frac{(1 - 1.2/8000)}{(1 - d_a/d_x)} = \frac{M_x^8 \times 0.999\,85}{(1 - d_a/d_x)} \qquad (4.6)$$

4.4.2.4 Buoyancy corrections

An important relationship is where an unknown mass of known density or volume is balanced against a standard weight of known conventional mass value under measured atmospheric conditions. In certain cases a buoyancy correction is necessary to determine the conventional value of the unknown mass. The correction δm is given by:

$$\delta m = m_s \left[\frac{1}{d_X} - \frac{1}{8000} \right] (d_a - 1.2) \qquad (4.7)$$

where:
m_s = conventional mass value of the standard weight
d_X = density of the unknown mass
d_a = density of air at the time of the weighing.

Example 1

A test mass of nominal value 500 g and density 7800 kg/m^3 is compared with a standard mass having a conventional mass value of 500.125 g in a laboratory with air density of 1.31 kg/m^3. Determine the buoyancy correction applicable to the conventional mass value of the test mass.

Applying the above equation we get:

$$\delta m = 500.125 \left[\frac{1}{7800} - \frac{1}{8000} \right] (1.31 - 1.2)$$

$\delta m = 0.000\,165$ g or 0.165 mg.

Example 2

In a chemical test laboratory a substance having a density of 4300 kg/m^3 is weighed on a weighing balance calibrated using standard masses having a density of 8000 kg/m^3 and conventional mass values referenced to standard air density, i.e. 1.2 kg/m^3. The mass indicated on the balance is 200.256 g.

Determine the buoyancy correction that should be applied to the weighing result.

Equation (4.7) could be used to estimate the buoyancy correction in this case as well. However, a knowledge of the air density of the environment in which the balance is located is required. Assuming the air density of the laboratory to be 1.31 kg/m^3 we proceed as follows:

$$\delta m = 200.256 \left[\frac{1}{4300} - \frac{1}{8000} \right] (1.31 - 1.2) = 0.002\,31 \text{ g} = 2.31 \text{ mg}$$

Thus 2.31 mg should be added to the weighing result of 200.256 g to obtain the corrected mass of the weighed material. If the required weighing uncertainty is less than ±0.001 g (1 mg) the buoyancy correction is definitely required.

4.5 Mass standards – types and classes

4.5.1 Types of masses

Masses are classified into categories depending upon their material and type of construction. The four main types are:

(a) Integral masses made from a non-magnetic stainless steel.
(b) Non-integral or two piece masses made from non-magnetic stainless steel. The mass value can be adjusted by the addition or removal of material from a small compartment usually underneath the screw knob.
(c) Masses made from brass (plated or unplated, integral or non-integral).
(d) Castiron masses, usually painted.

Masses of types 1 and 2 are used as reference standards for calibrating masses of lower accuracy classes and testing precision balances. For the verification of normal industrial and commercial weighing equipment masses of types 3 and 4 are adequate.

4.5.2 Classes of mass standards

The classification of mass standards into different classes is based on the maximum deviation of the 'conventional value' of the mass from its 'nominal value'. The OIML classification system is now widely used. The basic features of this system are given below.

4.5.2.1 OIML RI-111 classification

There are seven classes defined in the OIML RI-111: 1994 Recommendation, namely classes E1, E2, F1, F2, M1, M2 and M3. The exact requirements of the different classes are defined in the document. The applicable tolerances are indicated in Table 4.1.

The weights are made from a metal or metal alloy. Generally platinum iridium (class E1), stainless steel (classes E2, F1 and F2), brass or plated bronze (classes F2 and M1), cast iron (classes M2 and M3) are used. The

Table 4.1 Tolerances of OIML R111 masses. (Source: OIML)

Nominal value	Maximum permissible deviation in mg (±)						
	Class E1	Class E2	Class F1	Class F2	Class M1	Class M2	Class M3
50 kg	25	75	250	750	2500	7500	25000
20 kg	10	30	100	300	1000	3000	10000
10 kg	5	15	50	150	500	1500	5000
5 kg	2.5	7.5	25	75	250	750	2500
2 kg	1.0	3.0	10	30	100	300	1000
1 kg	0.5	1.5	5	15	50	150	500
500 g	0.25	0.75	2.5	7.5	25	75	250
200 g	0.10	0.30	1.0	3.0	10	30	100
100 g	0.05	0.15	0.5	1.5	5	15	50
50 g	0.030	0.10	0.30	1.0	3.0	10	30
20 g	0.025	0.080	0.25	0.8	2.5	8	25
10 g	0.020	0.060	0.20	0.6	2	6	20
5 g	0.015	0.050	0.15	0.5	1.5	5	15
2 g	0.012	0.040	0.12	0.4	1.2	4	12
1 g	0.010	0.030	0.10	0.3	1.0	3	10
500 mg	0.008	0.025	0.08	0.25	0.8	2.5	
200 mg	0.006	0.020	0.06	0.20	0.6	2.0	
100 mg	0.005	0.015	0.05	0.15	0.5	1.5	
50 mg	0.004	0.012	0.04	0.12	0.4		
20 mg	0.003	0.010	0.03	0.10	0.3		
10 mg	0.002	0.008	0.025	0.08	0.25		
5 mg	0.002	0.006	0.020	0.06	0.20		
2 mg	0.002	0.006	0.020	0.06	0.20		
1 mg	0.002	0.006	0.020	0.06	0.20		

(Reproduced from OIML RI 111-1994 with kind permission of the International Organization of Legal Metrology)

metal or alloy of classes E1, E2 and F1 weights must be practically non-magnetic.

The density of the material from which the weights are made should be such that a deviation of 10 per cent from the specified air density (1.2 kg/m^3) would not produce an error exceeding one-quarter of the maximum permissible deviation given in Table 4.1.

The metal or the alloy of which class M1 rectangular bar weights from 5 kg to 50 kg are made must be no more susceptible to corrosion and no more brittle than grey cast iron. Class M1 cylindrical weights up to 10 kg must be made of brass or of a material of quality at least equal to that of brass.

4.5.2.2 OIML RI-47 classification

Weights conforming to OIML International Recommendation RI-47: 1978 are used for the testing of high capacity weighing machines in accuracy classes III (medium) and IIII (ordinary). The applicable tolerances are given in Table 4.2. Weights conforming to RI-47 are in general made from grey cast iron.

Table 4.2 Tolerances for OIML RI 47 masses

Nominal value, kg	Maximum permissible relative error			
	3.3/10 000	1.7/10 000	1/10 000	0.5/10 000
	Corresponding absolute error, g			
50	17	8.5	5	2.5
100	33	17	10	5
200	66	33	20	10
500	170	85	50	25
1000	330	170	100	50
2000	660	330	200	100
5000	1700	850	500	250
Maximum no. of scale divisions	1000	3000	5000	10 000

(Reproduced from OIML RI 47-1979 with kind permission of the International Organization of Legal Metrology)

4.5.2.3 Other classifications

In the United States, the classifications of weights into classes and types are given in three primary publications:

ANSI/ASTM E 617-91 – Standard specification for laboratory weights and precision mass standards
NIST Handbook 105-1 (1990) – Specifications and tolerances for reference standards and field standard weights and measures.
NIST Handbook 44 – Specifications and tolerances for standard weights.

4.5.2.4 ASTM classification

Weights are divided into two types based on their design:

Type I – These weights are of one-piece construction and contain no added adjusting material.
Type II – Weights of this type can be of any appropriate design such as screw knob, ring or sealed plug. Adjusting material can be used as long as it is of a material at least as stable as the base material and is contained in such a way that it will not become separated from the weight.

The grade designations S, S′, O, P and Q describe design limitations such as density range of material, permitted surface area, surface finish, surface protection, magnetic properties, corrosion resistance and hardness.

Tables 4.3 and 4.4 indicate the tolerances for the different classes of ASTM weights.

Table 4.3 Tolerances for ANSI/ASTM E617 masses (Classes 1, 1.1 and 2)

Class	1		1.1	2	
Denomination	Individual mg	Group mg	Individual mg	Individual mg	Group mg
50 kg	125			250	
30 kg	75			150	
25 kg	62	135		125	270
20 kg	50			100	
10 kg	25			50	
5 kg	12			25	
3 kg	7.5			15	
2 kg	5.0	13		10	27
1 kg	2.5			5.0	
500 g	1.2			2.5	
300 g	0.75			1.5	
200 g	0.50	1.35		1.0	2.7
100 g	0.25			0.5	
50 g	0.12			0.25	
30 g	0.074			0.15	
20 g	0.074	0.16		0.10	0.29
10 g	0.050			0.074	
5 g	0.034			0.054	
3 g	0.034			0.054	
2 g	0.034	0.065		0.054	0.105
1 g	0.034			0.054	
500 mg	0.010		0.005	0.025	
300 mg	0.010		–	0.025	
200 mg	0.010	0.020	0.005	0.025	0.055
100 mg	0.010		0.005	0.025	
50 mg	0.010		0.005	0.014	
30 mg	0.010	0.020	–	0.014	0.034
20 mg	0.010		0.005	0.014	
10 mg	0.010		0.005	0.014	
5 mg	0.010		0.005	0.014	
3 mg	0.010		–	0.014	
2 mg	0.010	0.020	0.005	0.014	0.034
1 mg	0.010		0.005	0.014	

(Reproduced from ANSI/ASTM E 617 with the permission of the American Society for Testing and Materials)

4.5.2.5 NIST Handbook 105-1 classification

The tolerances specified in NIST Handbooks 105-1 and 44 are given in Tables 4.4 and 4.5.

Table 4.4 Tolerances for ANSI/ASTM E617 masses (Classes 3, 4, 5 and 6)

Class	3	4		5		6	
Denomination	mg	g	mg	g	mg	g	mg
5000 kg		100		250			
3000 kg		60		150			
2000 kg		40		100			
1000 kg		20		50			
500 kg		10		25		50	
300 kg		6.0		15		30	
200 kg		4.0		10		20	
100 kg		2.0		5		10	
50 kg	500	1.0		2.5		5	
30 kg	300		600	1.5		3	
25 kg	250		500	1.2		−	
20 kg	200		400	1.0		2	
10 kg	100		200		500	1	
5 kg	50		100		250		500
3 kg	30		60		150		300
2 kg	20		40		100		200
1 kg	10		20		50		100
500 g	5.0		10		30		50
300 g	3.0		6.0		20		30
200 g	2.0		4.0		15		20
100 g	1.0		2.0		9		10
50 g	0.60		1.2		5.6		7
30 g	0.45		0.90		4.0		5
20 g	0.35		0.70		3.0		3
10 g	0.25		0.50		2.0		2
5 g	0.18		0.36		1.3		2
3 g	0.15		0.30		0.95		2
2 g	0.13		0.26		0.75		2
1 g	0.10		0.20		0.50		2
500 mg	0.080		0.16		0.38		1
300 mg	0.070		0.14		0.30		1
200 mg	0.060		0.12		0.26		1
100 mg	0.050		0.10		0.20		1
50 mg	0.042		0.085		0.16		
30 mg	0.038		0.075		0.14		
20 mg	0.035		0.070		0.12		
10 mg	0.030		0.060		0.10		
5 mg	0.028		0.055		0.080		
3 mg	0.026		0.052		0.070		
2 mg	0.025		0.050		0.060		
1 mg	0.025		0.050		0.050		

(Reproduced from ANSI/ASTM E 617 with the permission of the American Society for Testing and Materials)

Table 4.5 Tolerances for NIST Handbooks 105-1 and 44 masses (5000–1 kg)

Denomination	NIST Handbook		
	105-1	44	
	Class F	Acceptance	Maintenance
kg	g	mg	mg
5000	500	–	–
3000	300	–	–
2000	200	–	–
1000	100	–	–
500	50	–	–
300	30	–	–
200	20	–	–
100	10	–	–
50	5.0	–	–
30	3.0	–	–
25	2.5	–	–
20	2.0	750	1500
10	1.0	500	1000
5	500	400	800
3	300	250	500
2	200	200	400
1	100	125	250

(Reproduced from NIST Handbooks 105-1 and 44 with the permission of the National Institute of Standards and Technology, United States Dept of Commerce)

4.5.2.6 Types and classes of balances

Weighing balances are classified into types by the design and weighing principle used and into classes by metrological criteria. Balances used for laboratory weighing and precise measurements are classified into the following main categories:

(a) Two-pan, three knife-edge balances.
(b) Single-pan, two knife-edge balances.
(c) Electromagnetic, force-compensation balances.

A brief description of each type follows.

4.5.2.7 Two-pan, three knife-edge balances

A schematic of a two-pan, three knife-edge undamped balance is shown in Fig. 4.1.

These balances consist of a main beam carrying a pan at each end. The beam has a central knife edge resting on a bearing pad when the balance is in operation. The two pans are supported by knife edges at the extremities of the beam. All the three knife edges lie in a plane. The central knife edge is nominally equidistant from the pan knife edges and due to this reason these balances are also known as equal arm balances. The balance beam is arrested

Table 4.6 Tolerances for NIST Handbooks 105-1 and 44 masses (500 g – 1 mg)

Denomination	105-1 Class F	NIST Handbook 44 Acceptance	Maintenance
g	mg	mg	mg
500	70	88	175
300	60	75	150
200	40	50	100
100	20	35	70
50	10	20	40
30	6.0	15	30
20	4.0	10	20
10	2.0	8	15
5	1.5	5	10
3	1.3	4	8
2	1.1	3	6
1	0.90	2	4
mg	mg	mg	mg
500	0.72	1.5	3.0
300	0.61	1.0	2.0
200	0.54	0.8	1.5
100	0.43	0.5	1.0
50	0.35	0.4	0.8
30	0.30	0.3	0.6
20	0.26	0.2	0.4
10	0.21	0.15	0.3
5	0.17	0.05	0.1
3	0.14	0.05	0.1
2	0.12	0.05	0.1
1	0.10	0.05	0.1

(Reproduced from NIST Handbooks 105-1 and 44 with the permission of the National Institute of Standards and Technology, United States Dept of Commerce)

(raised) while not in use and released (lowered) when a weighing is to be carried out. A pointer attached to the beam moving over a scale is used to read the rest point or the turning points of the beam. At present this type of balance is mostly used in high precision metrology laboratories for calibration of secondary and tertiary level mass standards.

There are two types of equal arm balances, undamped or free swinging balances and damped balances.

4.5.2.8 Undamped or free-swinging balance

Undamped balances are subject to only slight natural damping and come to rest after a long period of time. The rest point of these balances is determined by reading the turning points of the beam in oscillation. There are a number of formulae used to calculate the rest point. A widely used formula to calculate the rest point is:

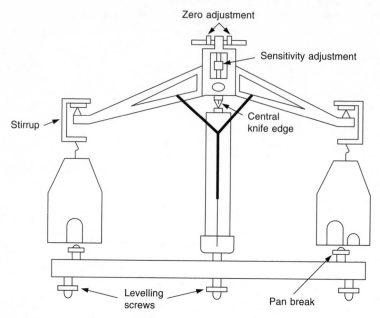

Figure 4.1 Two-pan, three knife-edge balance

$$\text{Rest point} = [(t_1 + t_3 + t_5)/3 + (t_2 + t_4)/2]/2 \qquad (4.8)$$

In the above equation t_1, t_2, t_3, t_4 and t_5 are the successive values of the turning points. Undamped balances are more sensitive than damped balances though they are not as convenient to use.

4.5.2.9 Damped balances

Damped balances have an arrangement to provide damping of the beam oscillations using air, oil or a magnetic field as the damping medium. Damping is generally arranged to be critical so that the pointer crosses the rest point once and comes to rest. The rest point of the balance is usually read off an optical scale fixed in front of the balance or in more modern balances an electronic digital display is provided.

4.5.2.10 Single-pan, two knife-edge balances

Both analytical and top-loading types of balances use this principle. A schematic diagram of an analytical balance is given in Fig. 4.2.

In this type of balance the beam carries a single pan at one of its ends. The beam has a knife edge fixed to it, which rests on a bearing. A counterweight is used to balance the beam when it is released on to the bearing. Two movable masses attached to screw pins are used to adjust the balance sensitivity and rest point. The beam is critically damped using an air dash-pot.

The balance also has built-in masses attached to the pan end of the beam.

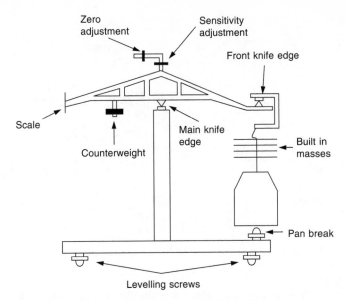

Figure 4.2 Single-pan, two knife-edge analytical balance

The masses can be lifted off the beam by a dial mechanism fixed to the balance case. Whenever a load, whose mass is to be determined, is placed on the pan, an equivalent mass is lifted off the beam by dialling, until the reading comes within the optical scale of the balance. The mass of the load placed is then read as the total sum of the dial readings plus the optical scale reading.

In the analytical type of balance the load is suspended below the balance beam and the beam is arrested during loading and unloading of the pan. For the top-loading balance the pan is supported above the beam by a parallelogram linkage and there is usually no arresting mechanism.

Since the mass to be supported by the knife edges is fairly constant, this type of balance is also known as a 'constant-load balance'. Also the sensitivity of the balance remains practically constant as the total beam load remains constant at all pan loads. Most older generation analytical balances are of this type.

4.5.2.11 Electromagnetic force-compensation balances

This type of balance works on the principle of electromagnetic force compensation and the gravitational force exerted on an object to be weighed is directly measured in contrast to comparison of forces done, in the case of beam type balances. Figure 4.3 shows a schematic diagram of the balance.

A coil rigidly attached to the pan linkage of the balance is placed in the annular gap of a magnet. When a load is placed on the pan a position sensor detects that the pan has been lowered and causes the current through the coil to be increased, generating a magnetic counterbalancing force bringing the

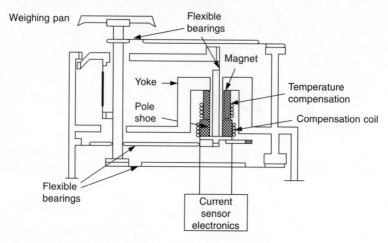

Figure 4.3 Electromagnetic force-compensation balance

balance pan to its original position. The compensating current is converted to a voltage by passing it through a resistor and read out on what is effectively a digital voltmeter.

In most of the modern electronically operated precision balances a microprocessor is incorporated in the circuitry. The level of sophistication of these instruments is very high, and there are many balance functions and features, which have to be learnt by a careful reading of the operation manual. Let us examine the electronic capabilities that are inherent to the basic balance and those that may be optional.

All balances have a digital display to indicate *mass*. Additional features include taring control, dual capacity and precision, selectable sampling period, inbuilt calibration mass, etc. A brief summary of these features follows.

Taring control

The taring control is a facility to zero the balance display when a load is placed on the pan. This facility permits one to subtract the mass of an object (such as a weighing boat or a watch glass) which is common to a series of weighings.

Dual capacity and precision

A feature that allows the balance capacity to be varied by a predetermined amount with a comparable change in precision.

Variable sampling period

Sampling period is the time interval used for averaging the sensing parameter.

Provision is made so that this interval can be varied. As the sampling interval is lengthened the overall balance weighing cycle is likewise lengthened.

Filters

Electronic filters that eliminate certain portions of the noise spectrum from the servo loop are provided.

Computer compatibility

Facilities to interface the balance output to an external computer using BCD (binary coded decimal), RS-232C or IEEE-488 interface protocols are provided.

Computation

A balance option that does computations such as counting, standard deviation calculations or user-defined computations.

Elimination of poor data

A facility to protect the user from collecting poor data, due to unusually strong air currents or vibrations, by not displaying the weighing results during these periods. Usually an override capability to cancel the protecting mechanism is provided.

Although the electromagnetic force compensation balances are widely used today, there are some occasions where the use of these balances may give rise to problems.

Weighing ferromagnetic material

The magnetic field associated with the servomotor may be changed by the presence of ferromagnetic materials giving rise to systematic errors in the indication of the balance. This can be verified by moving the material in and around the balance pan while observing the zero reading. The effect may be minimized by weighing below the balance pan if this is possible.

Electromagnetic radiation

The presence of a strong electromagnetic field may give rise to variable readings or malfunctioning.

Dust susceptibility

These balances are very sensitive to dust in the environment. When dust particles enter the gap between the permanent magnet and the pole pieces of the servomotor the precision and calibration of the balance can change. If the particles are ferromagnetic the balance may be rendered inoperable. Such environments should be avoided.

4.5.2.12 Mass comparators

Mass comparators are used for comparison of precise masses. Generally the construction of these balances is similar to the electromagnetic force-compensation type but they are built with more precision. Mass comparators are available in capacities in the range 2 g to 20 kg with excellent repeatability and linearity.

4.6 Industrial weighing systems

There is a large variety of equipment used for carrying out weighing operations in industrial situations. These systems use a variety of weighing principles. They are also classified as mechanical, electrical, hydraulic or pneumatic types. The most common types of weighing systems are operated mechanically or electrically.

4.6.1 Mechanical systems

Most mechanical weighers use knife-edge and lever systems. Some mechanical balances use pendulum systems to carry out the weighing operation. Mechanical systems often require no electrical power. Usually these are mass comparison devices independent of gravity and have very good accuracy. However, the maintenance requirement of these systems is high. Also they are not suitable for use in environments where deposits may form on the knife edges.

4.6.2 Electrical systems

Electrical weighing systems mainly consist of load cells connected either in series or parallel configuration. A load cell generates an electrical voltage output in proportion to the load applied on a fixed metallic member. Resistance strain gauges, capacitance, force balance and resonant wire cells are used as the load-sensing element in load cells.

There are many advantages of electrically operated systems. Electrical output is suitable for remote transmission and for interfacing with computers and modern instrumentation used in process control. A wide range of capacities (milligrams to thousands of tonnes) with high speed and accuracy are available. These systems are also suitable for both static and dynamic weighing, as well as weighing in hazardous areas provided special precautions are taken.

4.6.3 Pneumatic systems

In some industrial situations a pneumatically operated weighing system is used. A pneumatic weighing system is a force balance system, which uses air to transmit the force generated in a lever system. Usually compressed air is used as the force transmission medium.

Pneumatic systems require no electrical power and are intrinsically safe in hazardous environments. They can also be used for remote indication and are often fitted with self-indicating analogue dials. Accuracy of these systems is generally good.

On the debit side the maintenance level of these systems is high. They also need a clean compressed air supply.

4.6.4 Hydraulic systems

Hydraulic systems use fluids to transmit the force generated by the mass via a piston acting in a cylinder. These cylinders usually take the form of a load cell. Hydraulic systems require no electrical power supply. Tensile versions are available for suspended load. Generally maintenance level is high and fluid leakage is a problem. Also multiple cells and special hydraulic force summing devices are required for supported load. Hydraulic systems measure force and are therefore sensitive to gravitational variations. Accuracy of these systems is limited.

4.7 Accuracy classes of balances

A classification of weighing instruments in terms of their accuracy classes is given in OIML International Recommendation R76-1 for non-automatic weighing instruments. Non-automatic weighing instruments are classified into four accuracy classes, Special I, High II, Medium III, and Ordinary IIII. The classification is based on the verification scale interval (e) and the number of verification scale intervals (n).

4.7.1 Actual scale interval (d)

The actual scale interval (d) of a weighing instrument is the value expressed in units of mass of the difference between the values of two consecutive scale marks of an analogue instrument. In the case of an instrument with a digital display, it is the difference between two consecutive display readings.

4.7.2 Verification scale interval (e)

The verification scale interval is used for the classification and verification of a weighing instrument. In an instrument without an auxiliary indicating device, the verification scale interval is equal to the actual scale interval, i.e. $e = d$. For instruments with an auxiliary indicating device or non-graduated instruments the verification scale interval is determined in accordance with the requirements of Table 4.7.

Table 4.7 Accuracy classes of non-automatic weighing instruments

Accuracy class	Verification scale interval, e	Number of verification scale intervals,		Minimum capacity, min. (lower limit)
		Min.	Max.	
Special I	$0.001 \text{ g} \leq e$	50 000	–	$100e$
High II	$0.001 \text{ g} \leq e \leq 0.05 \text{ g}$	100	100 000	$20e$
	$0.1 \text{ g} \leq e$	5000	100 000	$50e$
Medium III	$0.1 \text{ g} \leq e \leq 2 \text{ g}$	100	10 000	$20e$
	$5 \text{ g} \leq e$	500	10 000	$20e$
Ordinary IIII	$5 \text{ g} \leq e$	100	1000	$10e$

(Reproduced from OIML RI 76-1-1992 with kind permission of the International Organization of Legal Metrology)

4.8 Calibration of balances

4.8.1 Precision balances

In general balances should be calibrated *in situ*, i.e. at the place where the balance is operated. This is especially applicable to balances, that measure gravitational force rather than compare masses. The stability of environmental conditions at the place of calibration is also important, especially the ambient temperature and humidity. Most precision balances are susceptible to changes in temperature and humidity, though this affects mostly the repeatability of the balance. Air currents and dust level in the atmosphere may also be influencing factors. Other influence factors of importance are strong magnetic or electromagnetic fields, which may affect mechanical components or electronics of the balance.

4.8.1.1 Reference standard

Balances are calibrated using standard masses of known conventional value and uncertainty. The combined uncertainty of the standard masses used for the calibration should at least be less than one-third of the specified or expected uncertainty of the balance under calibration. The masses should also have traceability to the international system with valid calibration certificates. The hierarchy of mass standards for balance calibration is given in Fig. 4.4.

4.8.1.2 Calibration parameters

The parameters of calibration depend on the type of balance. As the most common precision balance type in present day use is the direct reading electronic

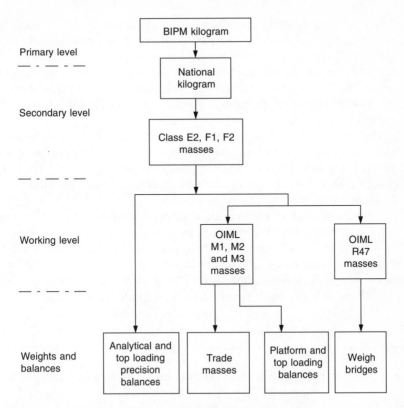

Figure 4.4 Calibration hierarchy for balances and masses

type, a brief outline of the calibration procedure of this type of balance is given.

4.8.2 Calibration of direct reading electronic precision balances

The main parameters that are tested in the process of calibration of this type of balance are scale value, linearity, repeatability, off-centre loading effect and hysteresis. To carry out the tests given in this section the balance should have reached equilibrium with its environment. This means that the balance should have been switched on for a considerable period of time before carrying out the tests. Usually an hour or two is added to the manufacturer's recommendation.

4.8.2.1 Setting scale value

Setting scale value is the process of establishing the correspondence between the mass units indicated in the display and the sensing parameter of the balance, i.e. electric current or voltage.

In some direct reading balances an inbuilt calibration mass is incorporated for this purpose and when the balance is switched on or when the calibration procedure is activated this mass is used automatically to set the scale value. In a majority of the balances an external mass close to the maximum capacity has to be used for setting the scale value. The procedure given in the technical manual of the balance should be followed in carrying out this operation.

The external mass has to be chosen carefully. The deviation of the mass from its nominal value should be less than half the discrimination of the balance if the scale value is to be set near the maximum capacity. If the scale value is to be set at a point less than its maximum capacity then the permissible deviation for the mass is reduced proportionately.

For example, a balance with a maximum capacity of 20 kg has a resolution of 0.1 g. The balance calibration is to be carried out using a mass of 20 kg nominal value. The deviation of the 20 kg mass should not exceed ±50 mg. If the operation is to be carried out at a 10 kg load, then the 10 kg mass used should have a deviation of less than ±25 mg. Reference to Table 4.1 will indicate that in both cases a mass of class E2 is required to set the scale value of the balance.

4.8.2.2 Repeatability

Repeatability is usually determined at half range and full range of the balance.

Procedure

Switch on and allow the balance to warm up for the time period recommended in the technical manual of the balance.

1. Choose a mass equal to approximately half the range of the balance.
2. Read zero and record, z_1.
3. Place the mass on the pan and record the reading, r_i, remove the mass.
4. Repeat steps 2 and 3 ten times and record the results.
5. Repeat steps 2, 3 and 4 using a mass approximately equal to the full range of the balance.

Tabulate results as shown in Table 4.8 and calculate the standard deviation (s) of the corrected readings from the equation:

$$s = \sqrt{\frac{\sum_{i=1}^{n} (r_i - \bar{r})}{n - 1}} \qquad (4.9)$$

where $\bar{r}$ is the mean value of the corrected readings, r_i, and n is the number of observations.

Table 4.8 Results of repeatability test

Number	Zero g	Reading g	Difference g
1	0.01	200.01	200.00
2	0.00	200.00	200.00
3	0.02	200.02	200.00
4	0.00	200.01	200.01
5	0.00	200.00	200.00
6	0.01	200.01	200.00
7	0.00	200.01	200.01
8	0.01	200.00	199.99
9	0.02	200.01	199.99
10	0.01	200.02	200.01
	Standard deviation, g		0.01

4.8.2.3 Linearity

Linearity is a measure of the deviation of the balance reading from the expected value. A set of calibrated masses with known conventional values and uncertainties is required to carry out the linearity test. A procedure often used is given below.

Procedure

The balance is tested at ten test points over its working range.

Switch on and allow the balance to warm up for the time period recommended in the technical manual of the balance. Divide the full range of the balance into ten equal steps, e.g. if the balance has a range of 200 g, each step will equal 20 g. Choose a set of weights having known values that can cover the full range of the balance. Proceed as follows:

1. Read and record the reading (zero).
2. Place a known mass M on the pan and record reading, r_1. Remove the mass.
3. Replace the mass and record the reading, r_2.
4. Remove the mass, record the reading (zero).
5. Repeat steps 2 to 4 for all the test points.
6. Tabulate and calculate results as shown in Table 4.9.

4.8.2.4 Off-centre loading effect

The off-centre load error occurs when the centre of mass of the object being weighed is off-centre on the pan. The test is done at the load recommended by the manufacturer of the balance (usually one-third or one-half the range of the balance).

Table 4.9 Results of linearity test

Pan load	Balance reading	Mean	Corr. balance reading	Value of standard mass	Correction
g	g	g	g	g	g
0	0.00				
20	20.01	0.00			
20	20.02	20.02	20.02	19.99	−0.03
0	0.00				
40	40.00	0.00			
40	40.01	40.01	40.01	39.99	−0.02
0	0.00				
60	60.02	0.00			
60	60.02	60.02	60.02	60.01	−0.01
0	0.00				
80	80.02	0.01			
80	80.01	80.02	80.01	80.01	0.00
0	0.01				
100	100.01	0.01			
100	100.00	100.01	100.00	99.99	−0.01
0	0.00				
120	120.02	0.00			
120	120.01	120.02	120.02	120.00	−0.02
0	0.00				
140	140.00	0.01			
140	140.01	140.01	140.00	139.99	−0.01
0	0.01				
160	160.01	0.01			
160	160.01	160.01	160.00	160.01	0.01
0	0.01				
180	180.00	0.01			
180	180.01	180.01	180.00	180.01	0.01
0	0.00				
200	200.00	0.00			
200	200.01	200.01	200.01	200.01	0.00
0	0.00				

Procedure

1. Switch on the balance and allow it to warm up for the time period recommended in the technical manual of the balance.
2. Choose a mass about one-third or one-half the full range of the balance.
3. For balances up to 10 kg use an aluminium or Perspex disc of 50 mm diameter. For balances of higher capacity the disc is not required.
4. Place the mass centrally on the disc (if used) and place the disc or the mass on the centre of the pan. Record the reading r_c.
5. Move the disc (or the mass) to the left, back, right and front edges of the pan and record readings r_l, r_b, r_r, and r_f respectively.

Calculate the maximum difference between the readings as the effect of off-centre loading as shown in Table 4.10.

Table 4.10 Results of off-centre loading test

Centre r_c	Front r_f	Back r_b	Left r_l	Right r_r	Maximum difference g
100.18	100.20	100.17	100.18	100.19	0.03

A mass of 100 g is placed on a disc of 50 mm diameter and moved to various positions on the pan. The balance readings obtained are given in the table.

4.8.2.5 Hysteresis

Procedure

1. Switch on the balance and allow it to warm up for the time period recommended in the technical manual of the balance.
2. Zero the balance and record the reading, z_1.
3. Place a mass M, equal to half the range on the pan, r_1.
4. Add extra mass to bring the balance reading close to full range.
5. Remove the extra mass and read the balance with M still on the pan, r_2.
6. Remove M and read and record zero, z_2.
7. Repeat the above procedure three times and average the differences $(r_1 - r_2)$ and $(z_1 - z_2)$.
8. Hysteresis of the balance = higher of the average of $(r_1 - r_2)$ or $(z_1 - z_2)$.

Table 4.11 Results of hysteresis test

Pan load		Half capacity of balance = 100 g		
		Run 1	Run 2	Run 3
Zero	z_1	0.00	0.00	0.00
M	r_1	100.03	100.03	100.03
$M + M'$		200.04	200.04	200.04
M	r_2	100.02	100.02	100.02
Zero	z_2	0.00	0.00	0.00
Calculations				
$z_1 - z_2$		0.00	0.00	0.00
$r_1 - r_2$		0.01	0.01	0.01

Bibliography

International and national standards

1. International Organization of Legal Metrology (OIML). International Recommendation R47-1979. Standard weights for testing high capacity weighing machines.
2. International Organization of Legal Metrology (OIML). International Recommendation R111-1994. Weights of classes E1, E2, F1, F2, M1, M2, M3.

3. ANSI/ASTM E617 (1991) Standards specifications for laboratory weights and precision mass standards. American National Standards Institution.
4. National Bureau of Standards. *NBS Handbook 110*. US Government Printing Office, Washington.
5. National Institute of Standards and Technology. *NIST Handbook 105-1*. US Government Printing Office, Washington.
6. National Institute of Standards and Technology. *NIST Handbook 44*. US Government Printing Office, Washington.
7. International Organization of Legal Metrology (OIML). International Recommendation R33-1979. Conventional value of the result of weighing in air.
8. International Organization of Legal Metrology (OIML). International Recommendation R74-1993. Electronic weighing instruments.
9. International Organization of Legal Metrology (OIML). International Recommendation R76-1-1992. Non-automatic weighing instruments, Part 1: Metrological and technical requirements – tests.
10. International Organization of Legal Metrology (OIML). International Document D11-1994. General requirements for electronic measuring instruments.

Introductory reading

1. Prowse, D.B. (1985) *Calibration of balances*. Commonwealth Scientific and Industrial Research Organization (CSIRO).
2. The Institute of Measurement and Control, London (1998) Guide to the measurement of mass and weight.
3. The Institute of Measurement and Control, London (1996) A code of practice for the calibration of industrial weighing systems.
4. National Bureau of Standards (1975) NBS Special Publication 420: The International Bureau of Weights and Measures 1875–1975. US Government Printing Office, Washington.

Advanced reading

1. National Bureau of Standards. NBS Monograph 133: Mass and mass values. US Government Printing Office, Washington.
2. National Bureau of Standards (1984) Special publication 700-1: A primer for mass metrology. US Government Printing Office, Washington.
3. *Electronic Weigh Systems Handbook* (1986) BLH Electronics.
4. Liptak, B. (1995) *Instrument Engineers' Handbook* CRC Press, LLC.
5. *Weighing and Force Measurement in the 90s* (1991) International Measurement Confederation (IMEKO).

5

Pressure measurements

5.1 Introduction

Pressure is one of the most important and widely measured quantities in industry. Accurate and traceable pressure measurement is important in many sectors, e.g. health, meteorology, transport, power. A variety of instruments are used for measurement of pressure and the basic operating principles of some of these instruments together with guidelines for their calibration are given in this chapter.

5.2 SI and other units

The SI unit for measurement of pressure is the newton (N) per square metre (m^2), having the special name pascal (Pa). The pascal is comparatively small in magnitude and therefore for most practical measurements decimal multiples kilopascal, megapascal and gigapascal are used. The other metric units commonly used for measurement of pressure are the bar, the kilogram-force per square centimetre, and millimetre of mercury. However, the use of the latter unit is strongly discouraged due to its inadequate definition.

A popular non-metric unit for measurement of pressure is the pound-force per square inch (lbf/in^2). In both metric and non-metric systems, sometimes pressures are erroneously indicated in mass units, e.g. '200 PSI', which should correctly be written as '200 pound-force per square inch' or '200 lbf/in^2'. The definitions, symbols and conversion factors of pressure units are given in Table 5.1.

The manometric units *millimetre of mercury* and *inch of water* depend on an assumed liquid density and acceleration due to gravity. Both of these assumptions inherently limit their relationship to the pascal. The use of these units, though given in the above table for sake of completeness, is strongly discouraged internationally. The pascal and its multiples and submultiples as appropriate to the magnitude of the pressure value are strongly recommended.

Table 5.1 Definitions, symbols and conversion factors of pressure units

Unit	Symbol	Definition	Conversion factor
pascal	Pa	$1 \ Pa = 1 \ N/1 \ m^2$	
kilogram-force per square centimetre	kgf/cm^2	$1 \ kgf/cm^2 = 1 \ kgf/1 \ cm^2$	$1 \ kgf/cm^2 = 98 \ 066.5 \ Pa$ (exactly)
bar	bar	$1 \ bar = 10^5 \ Pa$	$1 \ bar = 10^5 \ Pa$
pound-force per square inch	lbf/in^2	$1 \ lbf/in^2 = 1 \ lbf/1 \ in^2$	$1 \ lbf/in^2 = 6894.76 \ Pa$
millimetre of mercury	mmHg	See Note 1	$1 \ mmHg = 133.322 \ Pa$
inch of water	inH_2O	See Note 2	$1 \ inH_2O = 248.6 \ Pa$

Notes
1. The conventional millimetre of mercury is defined in terms of the pressure generated by a mercury column of unit length and of assigned density 13 595 kg/m^3 at 0°C under standard gravity of 9.806 65 m/s^2.
2. The conventional inch of water is defined in terms of the pressure generated by a water column of unit length and of assigned density 1000 kg/m^3 subjected to standard gravity of 9.806 65 m/s^2.

5.3 Absolute, gauge and differential pressure modes

If a vessel were to contain no molecules within it, the pressure would be zero. Pressures measured on the scale with zero pressure as the reference point are said to be *absolute pressures*.

The earth's atmosphere exerts a pressure on all objects on it. This pressure is known as the atmospheric pressure and is approximately equal to 100 kPa. Pressures measured in reference to the atmospheric pressure are known as *gauge pressures*. The difference between a pressure higher than atmospheric and atmospheric pressure is a positive gauge pressure, while the difference between atmospheric pressure and a pressure lower than atmospheric is referred to as negative gauge pressure or vacuum. Gauge pressure values being dependent on atmospheric pressure change slightly as the ambient pressure changes.

The relationship between absolute and gauge pressure is given below:

$$\text{Absolute pressure} = \text{gauge pressure} + \text{atmospheric pressure} \quad (5.1)$$

A differential pressure is the difference of pressure values at two distinct points in a system. For example, the flow of a fluid across a restriction in a pipe causes a pressure differential and this is used to determine the flow of the gas or liquid. This is the principle of the orifice plate as shown in Fig. 5.2.

5.4 Primary standards

A number of different physical standards are used for the realization of pressure values at the primary level. The two most common instruments for the positive

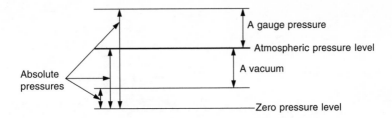

Figure 5.1 Pressure modes and their relationships

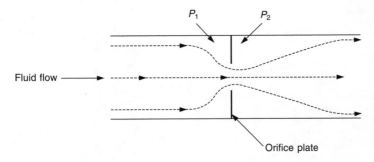

Figure 5.2 Differential pressure in an orifice plate

gauge pressure range are the mercury manometer and dead weight pressure tester. The spinning ball gauge standard is used in the negative gauge pressure (vacuum) range.

5.4.1 Mercury manometer

The basic principle of the mercury manometer is illustrated in Fig. 5.3. Vessels A and B are connected using a flexible tube. Vessel A is at a fixed level while vessel B can be moved up and down using a lead screw mechanism. The output pressure is obtained from vessel A.

A vacuum pump is sometimes used to evacuate the air above the meniscus of the moving vessel. Under these conditions the output pressure P_{out} is given by:

$$P_{out} = P_{in} + h \cdot \rho \cdot g \qquad (5.2)$$

where:

P_{in} = the pressure due to the gas above the meniscus of the moving vessel
h = the difference of height between the mercury levels of the two vessels
ρ = the density of mercury
g = local acceleration due to gravity.

In some designs, large diameter tubes (several tens of millimetres) are used to reduce capillary depression of the meniscus and other surface tension effects.

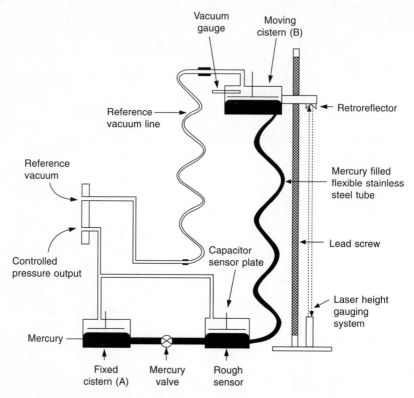

Figure 5.3 Mercury manometer

With these measures uncertainties in pressure of a few parts per million can be achieved. However the mercury temperature (typically to 0.005°C), the mercury density, the vertical distance between the mercury levels and the local value of gravitational acceleration have to be determined with low uncertainties. Individually built large-bore mercury manometers, using a variety of optical, capacitive, ultrasonic or inductive methods for detecting the mercury surface positions, are used in many national laboratories as primary standards. Slightly less capable instruments are available commercially and measure pressures up to about 3×10^5 Pa.

5.4.2 Dead weight pressure tester

A dead weight pressure tester consists of three main elements, namely the pressure balance, set of dead weights and the pressure source. A schematic of a dead weight pressure tester in its simplest form is shown in Fig. 5.4.

The pressure balance consists of a piston inserted into a closely fitting cylinder. The set of weights is usually made from non-magnetic stainless steel in the form of discs stackable on top of each other. Hydraulic pressure generated by a manual or electrically driven pump or pneumatic pressure

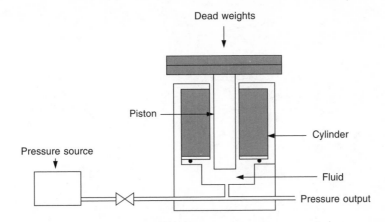

Figure 5.4 Dead weight pressure tester

obtained from a pressurized vessel is applied to the piston–cylinder assembly of the pressure balance.

Pressure testers used as primary level standards are calibrated by absolute methods by estimating the effective diameter and deformation characteristics of the piston–cylinder assembly together with the determination of the mass values of the weights (see Section 5.7.1.5). Dead weight pressure testers are used in the range 3 kPa (gas media, absolute or gauge mode) up to 1 GPa (hydraulic, gauge mode). Uncertainties of the order of ±0.001 per cent of the reading are attainable with these instruments.

5.5 Spinning ball gauge standard

A spinning ball gauge standard uses the principle of molecular drag to estimate the molecular density of a gas from which the pressure can be calculated. These standards can only be used for measurement of low absolute pressures below 10 kPa. The principle of the spinning ball gauge standard is illustrated in Fig. 5.5.

A ball made of magnetic steel a few millimetres in diameter is housed in a non-magnetic tube connected horizontally to a vacuum chamber. The ball is magnetically levitated and spun to few hundred revolutions per second using a rotating magnetic field. The driving field is then turned off and the relative deceleration of the ball is measured with magnetic sensors. The deceleration of the ball due to molecular drag is related through kinetic theory to molecular density and pressure of the gas. The lowest pressure that can be measured is limited by the residual drag caused by induced eddy currents.

An inert gas, usually dry nitrogen, is used as the pressure medium. The temperature of the tube is measured accurately using a calibrated thermocouple or other instrument. The spinning rotor gauge is used in the absolute pressure range 10^{-5} Pa to 10 Pa.

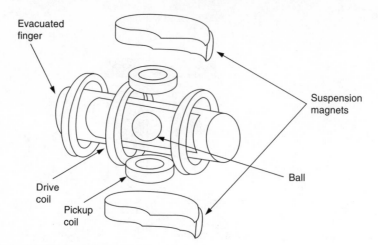

Figure 5.5 Spinning ball gauge standard

5.6 Secondary standards

The mercury manometer, dead weight tester and capacitance standard are the most commonly available secondary standards. A brief description of the capacitance pressure standard is given here as the other two standards, namely the mercury manometer and the dead weight tester, are covered in other sections.

5.6.1 Capacitance pressure standard

A schematic diagram of a capacitance pressure standard is shown in Fig. 5.6. Capacitance standards basically consist of a parallel plate capacitor whose plates are separated by a metallized diaphragm. The diaphragm and the two electrodes form two capacitors that are incorporated in an AC bridge circuit. The deflection of the diaphragm when a pressure is applied to one of the chambers is detected as a change in the capacitances. The two pressure chambers are electrically isolated and the dielectric properties are maintained constant.

The symmetrical design provides a more or less linear relationship between pressure and electrical output and differential pressures can be easily measured. To measure absolute pressures the reference chamber is evacuated.

Capacitance pressure standards operate in the pressure range 10^{-3} Pa to 10^7 Pa and generally have good repeatability, linearity and resolution. They also have high overpressure capability.

5.7 Working standards

The most commonly used working standard is the dead weight pressure tester.

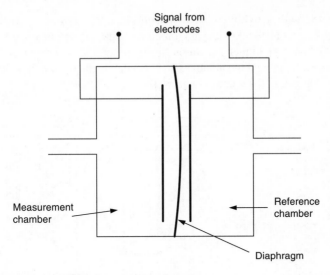

Figure 5.6 Capacitance pressure standard

A number of other instruments, such as precision bourdon or diaphragm type dial gauges, strain gauges, piezo resistive pressure sensors and liquid manometers are also used as working standards.

5.7.1 Dead weight pressure tester

5.7.1.1 The pressure balance

The most critical element of a dead weight pressure tester is the pressure balance. Pressure balances normally encountered are of two kinds, hydraulic (uses oil as the pressure medium) and pneumatic (uses air or nitrogen as the pressure medium). The latter type often has a facility to evacuate the ambient space around the piston–cylinder assembly, thus permitting their use for *absolute* as well as *gauge pressure* measurements.

The simple, re-entrant and the controlled clearance types shown in Fig. 5.7 are the three basic types of pressure balances in common use today. Although there are a number of technical and operational differences, the general principle of pressure measurement for the three types is the same.

5.7.1.2 Simple type

The geometry illustrated schematically in Fig. 5.7(a) is that of the simple type, where the piston and cylinder have basic cylindrical shapes. The calculation of the elastic deformation of this design is straightforward. Because fewer variables are needed to predict the deformation, the pressure coefficients can be estimated with a relatively small uncertainty. This design is commonly used for pressures up to 500 MPa and sometimes with appropriate modifications up to 800 MPa. At higher pressures distortion of the piston and cylinder becomes significant and the annular gap between the piston and cylinder is so large that the gauge does not operate well.

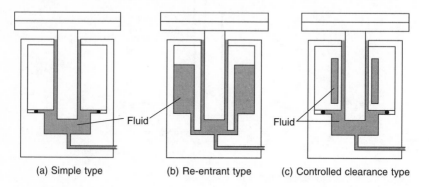

(a) Simple type   (b) Re-entrant type   (c) Controlled clearance type

Figure 5.7 Types of pressure balance

5.7.1.3 Re-entrant type

In the re-entrant type, illustrated schematically in Fig. 5.7(b), the pressure transmitting fluid acts not only on the base of the piston and along the engagement length of the piston and cylinder but also on the external surface of the cylinder. This external pressure reduces the gap between the piston and the cylinder thus reducing fluid leakage. The upper pressure limit is set by the reduction of the gap to an interference fit.

The disadvantage of this design is that it is difficult to accurately estimate the effects of distortion on the effective area of the piston and cylinder.

5.7.1.4 Controlled clearance type

In the controlled clearance type, illustrated schematically in Fig. 5.7(c), an external pressure is applied to the exterior surface of the cylinders enabling control of the gap between the piston and the cylinder. Using this design, in principle a very wide range of pressures can be covered using only one piston–cylinder assembly. However, in practice a series of assemblies is used to achieve the best sensitivity for a particular pressure range. This type of pressure balance is most commonly used in very high-pressure applications.

5.7.1.5 Simple theory of the pressure balance

The simple theory of the pressure balance is based on the application of laws of hydrodynamics or aerodynamics depending on whether the pressure transmitting medium is a liquid or a gas. The simple theory is explained using Fig. 5.8.

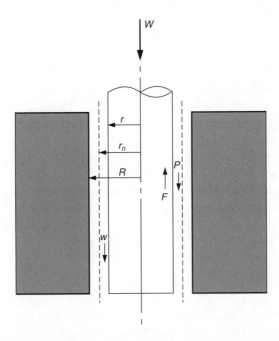

Figure 5.8 Diagrammatic representation of the pressure balance piston–cylinder assembly (with clearances greatly exaggerated)

The piston and cylinder are assumed to have straight and smooth cylindrical surfaces of circular cross-section of radii r and R respectively. The fluid pressure being measured, P_1, is applied to the base of the piston, while the top of the piston is exposed to ambient pressure, P_2. At the equilibrium condition the upward vertical force arising from the pressure difference $P_1 - P_2$ is balanced against a known downward gravitational force, W, which is applied to the piston by means of calibrated masses.

When the piston is in equilibrium:

$$W = \pi r^2 (P_1 - P_2) + F \tag{5.3}$$

where F represents a frictional force exerted on the vertical flanks of the piston by the fluid that is being forced to flow upwards under the influence of the pressure gradient.

The vertical component of the fluid velocity is at a maximum approximately halfway between the bounding surfaces (piston and cylinder surfaces) and it is zero at the bounding surfaces. The cylindrical surface at which the fluid velocity is maximum and frictional forces between adjacent layers are minimum is called the *neutral surface.* By equating the forces acting on the column of fluid of annular cross-section contained between the surface of the piston and the neutral surface, and denoting the downward force due to its weight as w the following equation is obtained:

$$w + F = \pi(r_n^2 - r^2)(P_1 - P_2) \tag{5.4}$$

Combining equations (5.3) and (5.4) gives:

$$P_1 - P_2 = \frac{w + W}{\pi \, r_n^2} \tag{5.5}$$

πr_n^2 is defined as the *effective area,* A_P, of the piston–cylinder assembly, i.e. the quantity by which the applied force must be divided to derive the applied pressure. It is a function of the dimensions of both the piston and the cylinder. The effective load strictly includes the force due to the mass of the annular column of fluid between the neutral surface and that of the true piston, but this is normally negligible. Hence, the applied gauge pressure $P = (P_1 - P_2)$, i.e. the amount by which the pressure within the system exceeds the external pressure at the reference level (the base of the piston or an identified plane), is to a good approximation given by:

$$P = \frac{W}{A_P} \tag{5.6}$$

In practice a number of deviations from the ideal form are found in both pistons and cylinders. Therefore, in order to calculate the effective area from dimensional data, measurements are required that yield information on the roundness and straightness of the components as well as their absolute diameters.

From the theory of elastic distortion it can be shown that the variation of the effective area, A_P, of a simple piston–cylinder with applied pressure P is essentially linear:

$$A_P = A_0(1 + aP) \tag{5.7}$$

where A_0 is the effective area for zero applied pressure, the deviations from linearity in practice being small. Also the dimensions of the components should be relatively large to reduce the uncertainties associated with diametral measurements to an acceptable level. Furthermore this method yields the value of the effective area at zero applied pressure and does not take into account the variation of the effective area of the assembly due to the elastic distortion of both piston and cylinder with applied pressure.

An estimate of the distortion coefficient, a, for the simple type of piston cylinder assembly can be calculated directly from dimensions and the elastic constants of the materials. However, due to the complexity of the forces acting on both components in any but the simplest designs, this method is somewhat limited.

For general purposes, these quantities are evaluated by comparing the pressure balance with a primary standard instrument, in a procedure often referred to as *cross-floating*, (see Section 5.9 on calibration of pressure measurement standards).

5.7.1.6 Corrections

In practice a number of corrections are required to determine gauge pressure using a dead weight pressure tester.

Temperature correction

The calibration certificate of a pressure tester will normally give the effective area value at a reference temperature of 20°C. If the temperature of the piston–cylinder assembly in use is different from the reference temperature a correction is required. This is usually combined with the pressure distortion coefficient a and expressed as:

$$A_{P,t} = A_{0,20}(1 + aP)[1 + (\alpha_p + \alpha_c)(t - 20)] \tag{5.8}$$

where:
$A_{p,t}$ = the effective area of the piston–cylinder assembly at applied pressure P and temperature t
$A_{0,20}$ = the effective area of the piston–cylinder assembly at zero applied pressure and 20°C
a = pressure distortion coefficient of the piston–cylinder assembly
α_p = linear expansion coefficient of the piston
α_c = linear expansion coefficient of the cylinder.

Evaluation of force

The general equation for the evaluation of downward force for an oil operated dead weight pressure tester is given by:

$$W = g\left[\sum\left\{M\left(1 - \frac{\rho_a}{\rho_m}\right)\right\} + B + H\right] + S \tag{5.9}$$

where:
W = net downward force exerted by the weights and the piston assembly
g = local acceleration due to gravity
M = conventional mass of the component parts of the load, including the piston
ρ_a = density of ambient air
ρ_m = density of the mass M, and can be significantly different for each load component
B = correction due to fluid buoyancy acting on the submerged parts of the piston
H = fluid head correction
S = correction for surface tension.

Air buoyancy correction

The factor $\left(1 - \dfrac{\rho_a}{\rho_m}\right)$ corrects for air buoyancy effects. It has a value of approximately 0.999 85 when working in air at one atmosphere and with steel weights. If there are significant differences of the densities of the weights, weight carrier and the piston, the corrections are separately worked out and added together.

The density of ambient air depends on the atmospheric pressure, temperature, relative humidity and composition. For very accurate work these parameters are measured and the air density calculated from an empirical equation. The approximate density of ambient air is 1.2 kg/m^3.

Fluid buoyancy correction

The fluid buoyancy force correction is calculated as an upthrust equivalent to weight of the volume of fluid displaced by the submerged part of the piston. The volume of fluid concerned depends on the reference level chosen for specifying the applied pressure.

Fluid head correction

The output pressure of a dead weight pressure tester is usually obtained at a level different from the reference level of the tester. A correction is then required to take account of the difference in levels. It is more convenient to combine the fluid head correction with the fluid buoyancy correction, and the combined correction factor expressed as a load correction is given by:

$$H + B = (hA - v)(\rho_f - \rho_a) \tag{5.10}$$

where:
h = difference in levels between the pressure output and reference plane of the pressure tester
A = nominal effective area of the piston
v = volume of fluid displaced by the piston
ρ_f = density of the fluid.

Surface tension effects

A correction to account for the surface tensional forces acting on the piston is included. This correction is given by:

$$S = s \cdot C \tag{5.11}$$

where:
S = force due to surface tension,
s = surface tension of the fluid,
C = circumference of the floating element at the point of emergence from the fluid.

Summary

Taking all the above correction terms into account, the applied pressure at the specified reference level is obtained from the equation:

$$P = \frac{g\left[\Sigma\left\{m\left(1 - \dfrac{\rho_a}{\rho_m}\right)\right\} + (hA - v)(\rho_f - \rho_a)\right] + s \cdot C}{A_{0,20}\,(1 + aP)[1 + (\alpha_p + \alpha_c)(t - 20)]} \qquad (5.12)$$

5.7.2 Portable pressure standard (pressure calibrator)

A variety of portable pressure standards, also known as pressure calibrators that use strain gauge, capacitance and piezo-resistive transducers, are available from a number of manufacturers. Usually these consist of a portable pressure pump, pressure transducer assembly and associated electronics and display. These are very convenient for calibration of pressure gauges and transmitters used on line in a large number of process industries. However, these instruments require frequent calibration against a secondary pressure standard maintained in a laboratory. Instruments ranging up to 800 kPa with an accuracy of ±0.5 per cent of the reading are available. Portable type dead weight pressure balances of the hydraulic type up to 70 MPa and pneumatic type up to 200 kPa are also in use.

5.8 Pressure measuring instruments

5.8.1 Liquid column instruments

5.8.1.1 Mercury barometers

Mercury barometers are generally used for measuring ambient pressure. There are two popular types, Fortin barometer and Kew pattern barometer.

Fortin barometer

A Fortin barometer (Fig. 5.9) can be used only to measure ambient pressure over the normal atmospheric pressure range. The height of the mercury column is measured using a vernier scale. A fiducial point mounted in the cistern determines the zero of the vertical scale. The mercury level in the cistern can be adjusted up or down by turning a screw to squeeze a leather bag. In making a measurement the instrument is made vertical with the help of a spirit level mounted on it and the screw is turned until the mercury meniscus in the cistern just touches the fiducial point. The vernier scale is then adjusted to coincide with the upper mercury meniscus and the reading is read off the scale.

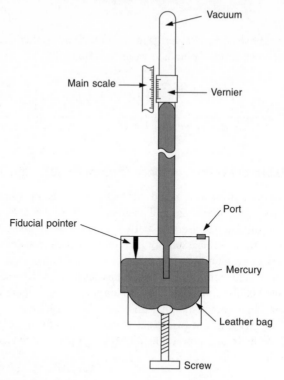

Figure 5.9 Fortin barometer

In addition to the corrections recorded in the calibration certificate, corrections are needed to take account of instrument temperature and value of gravitational acceleration. Details of these corrections are given in British Standard BS 2520.

Mercury barometers handled properly are very reliable instruments. They should be transported with extreme care.

Kew pattern barometer

A version of the Kew pattern barometer (Fig. 5.10) known as the station barometer, is similar to a Fortin barometer except that it has a fixed cistern. In this version the scale is contracted slightly in order to compensate for the varying mercury level in the cistern.

Kew pattern bench barometers are free standing and pressures from a few millibars to atmospheric can be measured. These use a pressure port and do not need total immersion calibration.

As in the case of the Fortin barometer, corrections are needed for changes in temperature and local gravitational acceleration. These are again given in the British Standard BS 2520.

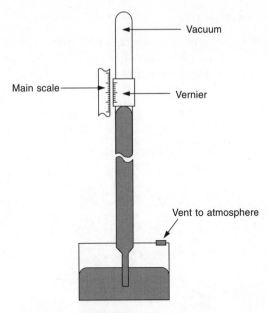

Figure 5.10 Kew pattern barometer

Precautions for handling of mercury barometers

Great care is needed in the transportation of mercury barometers primarily to avoid changing their metrological properties and exposing people and the environment to toxic mercury vapour. For transportation they should be sealed in rupture- and leak-proof plastic bags.

The glass tube of a Fortin barometer can be broken if mercury is allowed to oscillate up and down, while it is being moved in the upright position. To prevent this occurring or air entering the tube during transportation, the axial screw is turned until mercury has risen to within about 25 mm of the top of the tube. The barometer is then inclined slowly until mercury just touches the top of the tube, then continuing until the instrument is somewhere between horizontal and completely upside down.

Kew station barometers that do not have an axial screw should be treated similarly to Fortin barometers and turned slowly until horizontal or upside down.

In the case of Kew bench type barometers mercury in the tube should be isolated from the atmosphere before transportation, either with the tube nearly empty or nearly full. Some designs provide transportation sealing screws to achieve this but sealing the pressure port is sufficient. Additional packaging is applied between the tube and the barometer's frame, when transporting with the barometer's tube nearly full. The barometer is transported in the normal upright position.

Risk of spillage can also be reduced by ensuring that mercury barometers are placed in locations where they cannot be easily accidentally damaged.

5.8.1.2 U tube manometer

A U tube manometer is one of the most simple instruments used for measurement of pressure. It consists of a tube made from glass or other material (PVC or polythene) bent to the shape of a U and filled with a liquid, Fig. 5.11.

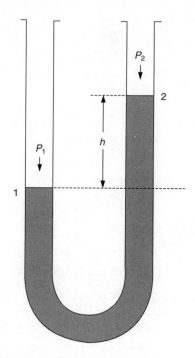

Figure 5.11 U tube manometer

The fundamental principle that the absolute pressure on the same horizontal plane in a given liquid is constant is used in the U tube manometer. In mathematical form this principle is expressed in the following equation:

$$P_1 = P_2 + h \cdot \rho \cdot g \qquad (5.13)$$

where:

P_1 and P_2 = pressure at points 1 and 2

$\quad h$ = the difference of height between the fluid levels of the two limbs

$\quad \rho$ = the density of manometric liquid

$\quad g$ = local acceleration due to gravity.

If $P_1 = P_2$, that is when both ends of the U tube are subjected to the same pressure, the levels of the liquid column will be at the same horizontal level. If, however, one limb is at a higher pressure than the other, the liquid column in that limb is depressed. The pressure difference between the two limbs is read off as the difference in heights of the liquid columns in the two limbs.

Mercury, water and oil are all used in various designs of manometer. For measuring large pressure differences, mercury is frequently used as its density is over 13 times greater than that of water or oil and thus, for a given pressure, it requires a much shorter column. Density of mercury is also considerably more stable than that of other liquids.

Water or oil liquid columns are used to measure low gauge and differential pressures. In some designs the manometer is inclined as this increases its sensitivity, the fluid having further to travel along the inclined column to achieve a given vertical movement. The traditional units for this type of measurement were *inches of water* or *millimetres of water,* but as units they are poorly defined and as mentioned earlier their continued use is strongly discouraged.

5.8.2 Mechanical deformation instruments

5.8.2.1 Bourdon tube gauge

A metallic tube of elliptical cross-section that is bent to form a circular arc is the sensing element of a Bourdon tube dial gauge. The application of a pressure to the open end of the tube straightens it out. The movement of the free end of the tube is amplified mechanically using gears and levers to operate a pointer. Bourdon tube dial gauges operate at pressures up to about 1.5 GPa and a typical mechanism is shown in Fig. 5.12.

Bourdon tube dial gauges are most commonly used for measuring gauge pressure but can also be used to measure absolute pressures by sealing the case. Differential pressure measurement is achieved by use of a second tube whose movement is mechanically subtracted from the main tube.

5.8.2.2 Diaphragm gauge

The diaphragm dial gauge (Fig. 5.13) is similar to a Bourdon tube dial gauge except that the moving element is a diaphragm. Its movement is transmitted through a connecting rod to an amplifying lever and gears that rotate a mechanical pointer.

Differential pressure is easily measured by applying it across the diaphragm.

5.8.2.3 LVDT

A linear variable differential transformer (LVDT) pressure transducer (Fig. 5.14) consists of a cylinder of ferromagnetic material moving inside a metallic tube. The end of the cylinder is attached to a deflecting component such as a diaphragm or bellows to which the test pressure is applied. Three coils are mounted on the tube. The central primary coil is excited with an alternating

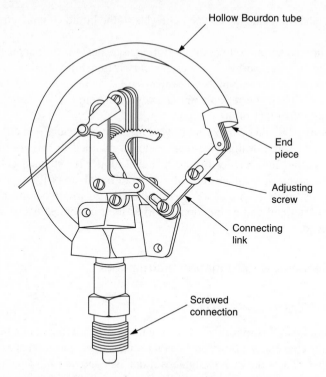

Hollow Bourdon tube

End
piece

Adjusting
screw

Connecting
link

Screwed
connection

Figure 5.12 Mechanism of a Bourdon tube pressure gauge

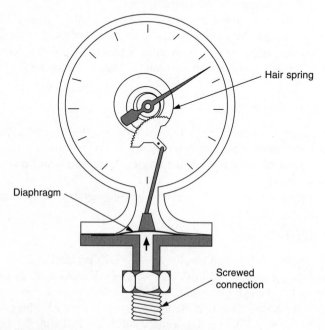

Hair spring

Diaphragm

Screwed
connection

Figure 5.13 Diaphragm dial gauge mechanism

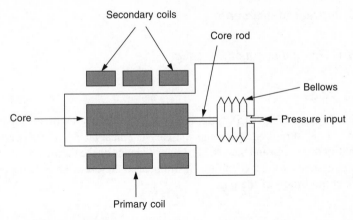

Figure 5.14 LVDT pressure transducer

voltage. The two sensing coils, one on either side are used for signal collection. As the magnetic cylinder moves within the tube, the magnetic field coupling, between the primary and secondary coils, changes. With suitable electronics, which may include temperature compensation, a linear relationship between cylinder position and output can be obtained. Sensors of this type are used in pressure transducers operating between pressures of about 10 mPa and 10 MPa.

The cylinder end may need support as the attachment of the pressure-sensing element increases the weight and stiffness of the LVDT. LVDT pressure transducers are more commonly available as gauge or differential pressure devices. Absolute pressure units are more complex.

5.8.2.4 Piezo electric devices

When certain types of crystalline materials are subjected to an external pressure, an electric charge proportional to the rate of change of applied pressure is generated on the crystal surface. A charge amplifier is used to integrate the electric charges to give a signal that is proportional to the applied pressure. The response is very fast, making these sensors suitable for dynamic pressure and peak pressure measurement. However, these sensors cannot be used for measurement of steady pressure values.

Early piezoelectric transducers used naturally grown quartz but today mostly artificial quartz is used. These devices are often known as *quartz pressure transducers*. A piezoelectric crystal being an active sensor requires no power supply. Also the deformation of the crystal being very small makes them have good high frequency response.

The major use of this type of sensor is in the measurement of very high frequency pressure variations (dynamic pressure) such as in measuring pressures in combustion chambers of engines. They are also capable of withstanding high overpressures.

5.8.3 Indirect instruments

5.8.3.1 Thermal conductivity gauges

Pirani gauge

In a Pirani gauge (Fig. 5.15), the energy transfer from a hot wire through a gas is used to measure the pressure of the gas. The heat energy is transferred to the gas by conduction and the rate of transfer depends on the thermal conductivity of the gas. The performance of these instruments therefore is strongly dependent on the composition of the gas.

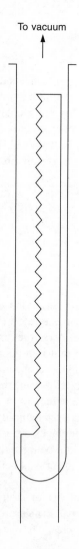

To vacuum

Figure 5.15 Pirani gauge

In the traditional configuration, a thin metal wire loop is sealed at one end of a glass tube whose other end is exposed to the gas. Tungsten, nickel, iridium or platinum is used as the material of the wire. In another type, the gauge sensor is a micro-machined structure, usually made from silicon covered by a thin metal film, such as platinum.

The wire or the metal film is electrically heated and its resistance, which is dependent on its temperature, is measured by incorporating the sensor element in a Wheatstone bridge circuit. There are three common operating methods: constant temperature method, constant voltage bridge and the constant current bridge.

The main drawback of Pirani gauges is their strong gas composition dependence and their limited accuracy. The reproducibility of Pirani gauges is usually fairly good as long as no heavy contamination occurs. The measuring range of Pirani gauges is approximately from 10^{-2} Pa to 10^5 Pa, but the best performance is usually obtained between about 10^{-1} Pa and 10^3 Pa. A variant of the Pirani gauge, known as *convection enhanced Pirani gauge*, is able to measure pressures in the range 10^{-2} Pa to 10^5 Pa.

5.8.3.2 Ionization gauges

A convenient method of measuring very low absolute pressures of a gas is to ionize the gas and measure the ionization current. Most practical vacuum gauges use electrons of moderate energies (50 eV–150 eV) to perform the ionization. The resulting ion current is directly related to pressure and a calibration is performed to relate the gas pressure to ionization current. However, these can only be used over a finite range of pressures. The upper pressure limit is reached when the gas density is so large that when an ion is created it has a significant probability of interacting with either a neutral gas molecule or free electrons in the gas so that the ion is itself neutralized and cannot reach the collector. For practical purposes this can be taken as 10^{-1} Pa. The lower pressure limit of an ionization gauge is around 10^{-6} Pa. This limit is reached when either electric leakage currents in the gauge measuring electronics become comparable to the ion current being measured or when another physical influence factor (e.g. extraneous radiation) gives rise to currents of similar magnitude.

Two types of ionization gauges are in widespread use, the *hot cathode* ionization gauge and the *cold cathode* ionization gauge.

Triode gauge

The triode gauge is a hot cathode type gauge. The gauge has been originally developed from the electronic valve. Electrons are emitted from a hot filament along the axis of the cylindrical grid, Fig. 5.16. The ions are created mainly inside the grid and are attracted to the cylindrical anode around the grid. The usual pressure range of the instrument is about 10^{-1} Pa to 10^{-6} Pa. A special

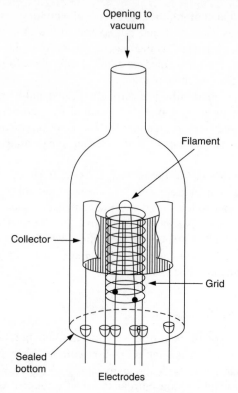

Figure 5.16 Triode gauge

design, the Schultz–Phelps gauge, can operate in the approximate range 10^2 Pa to 10^{-2} Pa.

Bayard–Alpert gauge

In the Bayard–Alpert design, the hot filament is outside of the cylindrical grid, Fig. 5.17. Ions are created mainly inside the grid and are collected on an axial collector wire. Some of the electrons produced as a result of the ionization of the gas molecules will generate X-rays when they hit the grid. X-rays hitting the collector may eject electrons from the surface and they will be indistinguishable from ions arriving at the collector. Due to the much smaller solid angle subtended by the collector wire fewer of the X-rays will strike the collector, resulting in a significantly lower pressure limit than for the triode gauge. This is the most common configuration for a hot filament ionization gauge. The pressure range is roughly 10^{-1} Pa to 10^{-9} Pa.

Penning gauge

The Penning gauge is a cold cathode type gauge. A schematic of the gauge head is shown in Fig. 5.18. In this gauge both electric and magnetic fields are

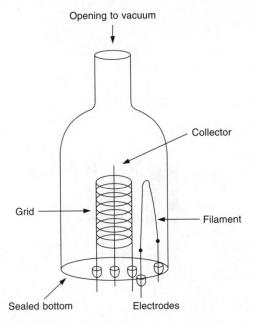

Figure 5.17 Bayard–Alpert gauge

used to generate and collect the ions. The anode may take the form of a ring or cylinder. When the electric field is high enough (a few kV DC), a gas discharge is initiated by the use of a miniature ultraviolet light source. Emission of electrons then takes place from the cathode plates. The loop anode collects ions. The pressure range is approximately 10^{-1} Pa to 10^{-7} Pa.

5.9 Calibration of pressure standards and instruments

5.9.1 General considerations

The most important general concepts relating to calibration of pressure standards and measuring instruments are discussed in this section.

5.9.1.1 Reference standard

The hierarchy of pressure measurement standards is given in Fig. 5.19, which may be used for selection of the next higher level standard for the calibration of a particular standard or instrument. However, the traceability path given in this diagram is not the only possible solution. What is indicated is one of many possible solutions.

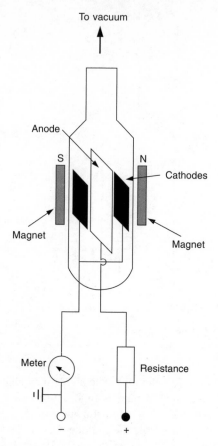

Figure 5.18 Penning gauge

Primary standards are the dead weight pressure tester, mercury manometer and spinning ball gauge. Secondary standards are those calibrated against the primary standard, namely dead weight pressure testers and capacitance standards. Working standards are dead weight testers, precision dial gauges (Bourdon tube or diaphragm type) and portable field type standards. A variety of these standards of different types (dead weight pressure balances, piezo-resistive devices, strain gauge type) are available. These are very useful for field calibrations.

5.9.1.2 Test uncertainty ratio

The uncertainty required for a particular standard or test instrument is an important criterion to consider before a calibration is performed. The uncertainty assigned to a primary, secondary or working standard is usually quoted at 95 per cent confidence level or coverage factor $k = 2$. However, this is only one component of the uncertainty budget. All other components of the system

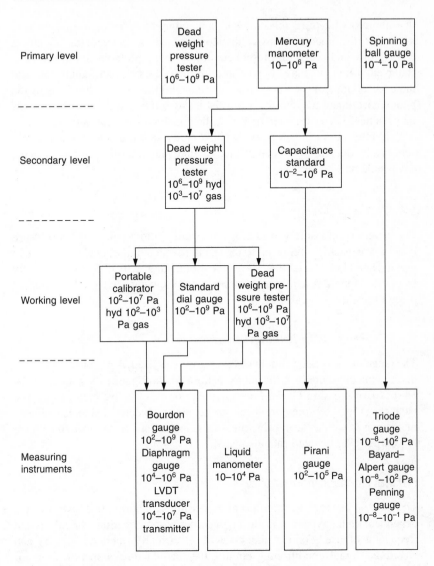

Figure 5.19 Hierarchy of pressure measurement standards

should be estimated. The usual criterion is that the combined uncertainty of the pressure calibration system including the higher level standard used should be at least three to four times smaller than the required uncertainty of the device under calibration. In some circumstances, when the item to be calibrated has a very low uncertainty a test uncertainty ratio of 1:1 may have to be used.

5.9.1.3 Reference conditions

Primary, secondary and working level pressure standards are always calibrated in a laboratory having stable ambient temperature and pressure. Most pressure measuring instruments are also calibrated in an environment with stable ambient temperature and pressure. In field conditions using portable equipment such stable environments may not be found. In such cases it is difficult to evaluate the uncertainties of the calibration in a meaningful way as all effects of the poor environment may not be quantifiable. Also the reference standard itself may not have been calibrated under similar conditions. It is thus more common to calibrate measuring instruments also under stable conditions and apply separately determined corrections to take account of poor in-service environments.

5.9.1.4 Local gravity

The generated pressure values given in a calibration certificate of a pressure balance are usually referenced to the standard value of the acceleration due to gravity, namely 9.806 65 m/s^2. If the local value of gravity differs significantly from the standard gravity the generated pressure values need to be corrected using the value of local gravity.

5.9.1.5 Range of calibration

The range of calibration should be carefully considered. A pressure measuring instrument or standard is normally calibrated throughout its total range at least at five pressure levels. For determination of effective area of pressure balances at least ten pressure levels are required. To detect hysteresis effects, increasing as well as decreasing pressure is used. Calibration over 20 per cent to 100 per cent of rated range is usual.

5.9.1.6 Recalibration interval

The interval between calibrations of a pressure standard or test instrument is dependent on the type of transducer, uncertainty specification, conditions and frequency of use. For instruments with electrically operated sensors and electronic signal conditioning circuits the manufacturer's specification for the device, particularly the specification for the long-term stability, is a good starting point for the determination of the recalibration interval. In most standards laboratories secondary standard dead weight testers (or pressure balances) are given a three-year recalibration interval. Diaphragm capacitance standards and associated electronics are susceptible to drift and may need a shorter interval (one to two years).

   It is possible to determine the approximate recalibration interval after several calibrations, if *as found* calibration data are recorded in the calibration report. If the calibration values change significantly at each calibration, then the interval may be too long. On the other hand if there is no appreciable change in the calibration values, the interval may be too short.

5.9.1.7 Pipework and tubing

Most positive gauge pressure measuring instruments are calibrated by connecting their pressure ports to a pressure standard using suitable pipework or special tubing rated for high pressure work. It is important to make sure that the pipework or tubing used is of good quality, undamaged and has a rating higher than the maximum pressure to be applied. The piping and joints should be inspected for leaks before the application of the maximum pressure. It is very important to ensure that the system is safe for operation at the maximum pressure envisaged. Guidance is available from a number of sources. Particularly the *High Pressure Safety Code* published by the High Pressure Technology Association.

5.9.1.8 Pressure medium

It is necessary to use the same pressure medium for the calibration as when the instrument is used. For example, instruments for measuring gas pressure should not be calibrated using oil as the pressure medium. The reverse is also true. In circumstances where two different media have to be used a separator cell is used to transfer the pressure from one medium to the other.

The pressure medium used, whether it be oil or a gas, should be clean and free of moisture. Filtered air or dry nitrogen is the preferred gas for calibration of gas pressure measuring instruments. Mineral or synthetic oils are used for hydraulic instruments. Certain oils may not be compatible with materials of some pressure system components. Also electrical conductivity of the oil is important when resistance gauges are being used. Instrument manufacturers usually recommend commercial oil types, which are suitable and their advice should be followed.

5.9.1.9 Instrument adjustment

Calibration of an instrument necessarily involves adjustment of the instrument where this is possible. Adjustments are performed until the deviations are minimum throughout the useful range. A number of repeat adjustments and test runs are required before an optimum level of deviations can be obtained.

5.9.2 Calibration of working standard dead weight pressure testers

Generally, working standard dead weight pressure testers are calibrated by comparing them with a secondary standard dead weight pressure tester in a procedure known as cross floating. Two methods are possible:

(a) Method A – Generated pressure method

The deviation of the nominal pressure value (usually marked on the weights) from the generated pressure is determined at a number of loads. The repeatability of the pressure balance is also determined. The determination of the conventional mass values of the weights and other floating components is optional.

(b) Method B – Effective area determination method

The following are determined:

(i)   The conventional mass values of all the weights, weight carrier and the piston of the pressure balance if removable.
(ii) The effective area A_P of the piston–cylinder assembly of the pressure balance as a function of pressure, at the reference temperature (usually 20°C).
(iii) The repeatability as a function of the measured pressure.

In method A, the deviation of the nominal pressure from the generated pressure and its uncertainty at each pressure level is ascertained. Method B, which is more time consuming, produces a complete calibration certificate with values for the effective area, mass values of the weights and their uncertainties.

The choice of the procedure to be followed in a particular case depends on a number of considerations, the most important being the uncertainty of the instruments to be calibrated using the dead weight tester.

5.9.2.1 Cross floating

In a cross floating system two pressure balances are interconnected into a common pressure system. When the loads on the two pressure balances are adjusted so that both are in equilibrium, the ratio of their total loads represents the ratio of the two effective areas at that pressure. The attainment of the equilibrium condition is the critical part of the measurement.

A single medium hydraulic cross float system between two dead weight pressure testers is shown in Fig. 5.20. A variable volume pump is used to pressurize the system and to adjust the float positions of the two pressure balances. An isolation valve is used to isolate the pump from the system to check for leaks. The two pressure balances are isolated from each other for determining sink rates of each balance independently. A sensitive differential pressure indicator and a bypass valve are inserted in line between the two pressure testers. The differential pressure indicator, though not essential, serves a useful purpose by indicating the pressure difference between the two testers and speeds up obtaining a balance.

A similar arrangement is used in a two media (hydraulic and gas) cross float system except that an additional component, a media separator cell, is required to transmit the pressure from one medium to the other. A differential pressure indicator is also used, as in the case of the single medium system.

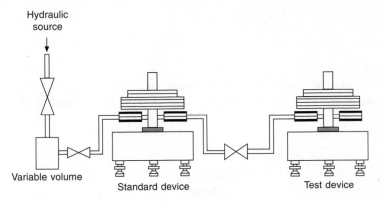

Figure 5.20 Typical arrangement for cross floating two hydraulic dead weight pressure testers

5.9.2.2 Estimation of uncertainty

The combined uncertainty of the measured pressure depends on a number of input uncertainties (see Chapter 9). The combined standard uncertainty, and the expanded uncertainty are calculated in conformity with the procedure given in Chapter 9, using the standard uncertainties estimated for each input component.

The main input components for method A and method B are listed below.

Method A

Type A uncertainties

Repeatability of the pressure balance is estimated at all loads by computing the standard deviation of the difference between the nominal and generated pressure. It is expressed in pascals, or as percentage of the nominal pressure.

Type B uncertainties

(a) Uncertainty of the pressure reference standard.
(b) Uncertainty of the mass values.
(c) Uncertainty of the local gravity.
(d) Uncertainty due to temperature.
(e) Uncertainty due to the head correction.
(f) Uncertainty due to tilt (negligible if perpendicularity was duly checked).
(g) Uncertainty due to air buoyancy, if significant.
(h) Uncertainty due to spin rate and/or direction, eventually.
(j) Uncertainty of the residual pressure (absolute mode only).

Method B

Type A uncertainty

Repeatability of the pressure balance is estimated at all loads by computing the standard deviation of the difference between the reference and generated pressure. It is expressed in pascals, or as percentage of the reference pressure.

Type B uncertainties

(a) Uncertainty of the mass values.
(b) Uncertainty of the measured effective area, including the uncertainty estimated using a type A method.
(c) Uncertainty due to the pressure distortion coefficient, when relevant, including the uncertainty estimated using a type A method.
(d) Uncertainty of the local gravity.
(e) Uncertainty due to the temperature of the balance.
(f) Uncertainty due to the air buoyancy.
(g) Uncertainty due to the head correction.
(h) Uncertainty due to tilt (negligible if perpendicularity was duly checked).
(j) Uncertainty due to spin rate and/or direction, eventually.
(k) Uncertainty of the residual pressure (absolute mode only).

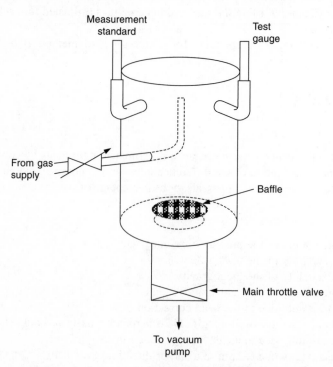

Figure 5.21 Configuration for vacuum gauge calibration

5.9.3 Calibration of vacuum gauges

Vacuum gauges are calibrated by connecting them to a sufficiently large vacuum chamber. A general configuration used is shown in Fig. 5.21.

The chamber is used only for calibration purposes and is kept as clean as possible. The vacuum is generated by an oil diffusion or turbo molecular pump backed up by a rotary pump. A throttle valve is used to connect the vacuum pump to the chamber. High vacuum isolation valves are used to connect the test instrument and the measurement standard. A clean gas source (nitrogen) connected through a needle valve allows a small amount of gas to be admitted to obtain different pressure levels. It is possible to obtain an equilibrium constant pressure in the chamber by adjustment of the throttle valve and the needle valve. Pressure indicated by the test instrument and the standard is recorded for each pressure level. Several repeat runs are carried out.

Bibliography

International and national standards

1. ISO 3529-1 1981 Vacuum Technology – Vocabulary – part 1: General terms. International Organization for Standardization.
2. ISO 3529-3 1981 Vacuum Technology – Vocabulary – part 3: Vacuum gauges. International Organization for Standardization.
3. BS 2520: 1983 British Standard – Barometer conventions and tables, their application and use. British Standards Institution.
4. BS 6134: 1991 British Standard – Specification for pressure and vacuum switches. British Standards Institution.
5. BS 6174: 1992 British Standard – Specification for differential pressure transmitters with electrical outputs. British Standards Institution.
6. BS 6739: 1986 British Standard – Code of practice for instrumentation in process control systems: installation, design and use. British Standards Institution.
7. ANSI/ASHRAE 41.3. 1989. Method for pressure measurement. American National Standards Institute.
8. ANSI/ASME PTC 19.2. 1987. Pressure measurement instruments and apparatus – part 2. American National Standards Institute.
9. Heydemann, P.L.M. and Welch, B.E. (1975) NBS Monograph, *Part 3, Piston Gauges*. National Bureau of Standards.
10. International Recommendation No. 110 – Pressure Balances – General. International Organization of Legal Metrology.
11. RISP-4. (1998) Dead weight pressure gauges. National Conference of Standards Laboratories.

Introductory reading

1. *Guide to Measurement of Pressure and Vacuum* (1998) Institute of Measurement and Control.
2. Chambers, A., Fitch, R.X. and Halliday, B.S. (1989) *Basic Vacuum Technology*, IoP Publishing, Adam Hilger.

3. Harris, N. (1989) *Modern Vacuum Practice*, McGraw-Hill.
4. Cox, B.G. and Saville, G. (eds) (1975) *High Pressure Safety Code*, High Pressure Technology Association.
5. Hucknall, D.J. (1991) *Vacuum Technology and Applications*, Butterworth-Heinemann, Oxford.
6. Lewis, S.L. and Peggs, G.N. (1992) *The Pressure Balance – A Practical Guide to Its Use*. HMSO.
7. Noltingk, B.E. (ed.) (1995) *Instrumentation*, 2nd edn, Butterworth-Heinemann, Oxford.
8. O'Hanlon, L.F. (1989) *A User's Guide to Vacuum Technology*, 2nd edn, Wiley.

Advanced reading

1. Berman, A. (1985) *Total Pressure Measurements in Vacuum Technology*, Academic Press.
2. Dadson, R.S., Lewis, S.L. and Peggs, G.N. (1982) *The Pressure Balance – Theory and Practice*, HMSO.
3. Herceg, E.E. (1976) *Handbook of Measurement and Control (Theory and Application of the LVDT)*, Schaevitz Engineering.
4. Leck, J.H. (1989) *Total and Partial Pressure Measurements in Vacuum Systems*, Blackie & Son.
5. Pavese, F. and Molinar, G. (1992) Modern Gas-Based Temperature and Pressure Measurements. International Cryogenic Monograph Series. Timmerhaus, K.D., Clark, A.F. and Rizzuto, C. (eds) Plenum Press.
6. Peggs, G.N. (ed.) (1983) *High Pressure Measurement Techniques*. Applied Science.
7. Wutz, W., Adam, H. and Walcher, W. (1989) *Theory and Practice of Vacuum Technology*, Vieweg.
8. Fitzgerald, M.P. and McIlmith, A.H. (1993/94) Analysis of Piston–Cylinder Systems and the Calculation of Effective Areas, *Metrologia*, 30, 631–634.

6

Measurement of force

6.1 Introduction

Force measuring instruments such as load cells, proving rings, universal testers and compression testers are used in a number of industries. A load cell transducer is the principal component in a number of industrial type weighing machines. Testing laboratories in the construction industry are heavily dependent on accurate and traceable tensile and compressive strength testers. Force measurement is performed in a number of other sectors as well, e.g. manufacture of cables, power generation, power transmission and transport.

6.2 SI and other units

The SI unit for measurement of force is the newton. It is defined as the force required to accelerate a mass of 1 kg through 1 m/s^2. The newton is comparatively small in magnitude and therefore for most practical measurements decimal multiples, kilonewton, meganewton and giganewton are, used.

The kilogram-force and tonne-force are the other metric units commonly used for measurement of force. The kilogram force is the force experienced by a mass of one kilogram due to an acceleration of 9.806 65 m/s^2. The acceleration of 9.806 65 m/s^2 is known as standard acceleration and was introduced in order to define force units independent of the acceleration due to gravity, yet approximately equal to gravitational units. The tonne-force is equal to one thousand kilogram force, which is the same force required to accelerate a mass of 1000 kilograms through the standard acceleration (9.806 65 m/s^2).

In non-metric units the pound-force, defined as the force required to accelerate a mass of one pound through the standard acceleration or the ton-force, which is 2240 pound-force, is used. In both SI and non-metric systems, sometimes force or load values are erroneously indicated in mass units e.g. '100 tonne load', which should correctly be written as 'a load of 100 tonne-force' or 'a load of 100 tf'. A summary of the force units is given in Table 6.1.

Table 6.1 Summary of the force units

SI units	Other units	Symbol	Definition
newton		N	$1 \text{ N} = 1 \text{ kg} \times 1 \text{ m/s}^2$
kilonewton		kN	$1 \text{ kN} = 1000 \text{ N}$
meganewton		MN	$1 \text{ MN} = 1\ 000\ 000 \text{ N}$
	kilogram-force	kgf	$1 \text{ kgf} = 1 \text{ kg} \times 9.806\ 65 \text{ m/s}^2$
	tonne-force	tf	$1 \text{ tf} = 1000 \text{ kgf}$
	pound-force	lbf	$1 \text{ lbf} = 4.448\ 22 \text{ N}$
	ton-force	tonf	$1 \text{ tonf} = 9.964\ 02 \text{ kN}$

6.3 Primary standard

The primary standard of force measurement is the dead weight force standard machine. There are several types of these machines. The basic principle of the dead weight force standard machine is illustrated in Fig. 6.1. In this standard, the gravitational force exerted on a set of masses is utilized to generate the force. The force F generated by a mass M installed in a force standard machine under gravitational acceleration g is given by the following equation:

$$F = Mg \left[1 - \frac{\rho_a}{\rho_M} \right] \qquad (6.1)$$

The factor within the square bracket corrects for the air buoyancy effect. ρ_a and ρ_M are the densities of air and the mass respectively. It can be seen from equation (6.1) that to determine F an accurate value for the acceleration due to gravity (g) is required.

6.4 Secondary standards

6.4.1 Lever or hydraulic force standard machines

Higher capacity force standard machines generally use a hydraulic or mechanical lever system to increase the force generated by a set of dead weights. These machines are classified as secondary force standards as they require calibration against a primary standard. Usually a number of load cells in parallel are used as the transfer standard. Forces as high as 20 MN have been generated, using these machines. The schematic of a hydraulic force standard machine is given in Fig. 6.2.

6.4.2 Proving ring

A proving ring is a ring made from alloy steel. The relationship between the central deflection of the ring and the applied load at a specified temperature is used as the force standard. In a good quality proving ring this relationship remains unchanged for a considerable period of time if the ring has not been subjected to overloading, shock or any other deleterious effect. A proving ring can be used in either compression or tensile modes. The principle of operation of a proving ring is illustrated in Fig. 6.3(a).

A proving ring has to be calibrated in a dead weight force standard machine to establish the force–deflection relationship. When carefully used proving rings are excellent secondary standards.

Secondary standard proving rings up to 2 MN capacity having relative uncertainties of the order of ±0.01 per cent of full scale are commercially available.

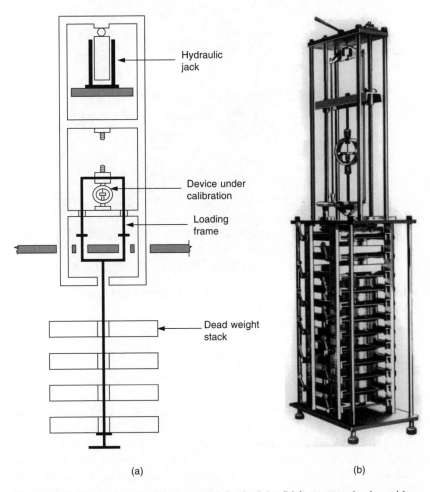

Hydraulic jack

Device under calibration

Loading frame

Dead weight stack

(a)   (b)

Figure 6.1 Dead weight force standard: (a) principle; (b) force standard machine of 5000 N capacity. (Source: Morehouse Instrument Co., USA)

Figure 6.2 20 MN capacity hydraulic force standard machine.
(Source: National Metrology Institute of Japan)

6.4.3 Load cell

Although load cells are commonly used for force measurement in a number
of instruments, such as universal testing machines, weigh bridges, cable testers,
etc., those having the required uncertainty and stability for use as secondary
standards are also available from a number of manufacturers. The principle of
operation of load cells is given in the section on force measuring instruments.
Secondary standard load cells up to 5 MN having relative uncertainties of
±0.05 per cent of full scale are available commercially.

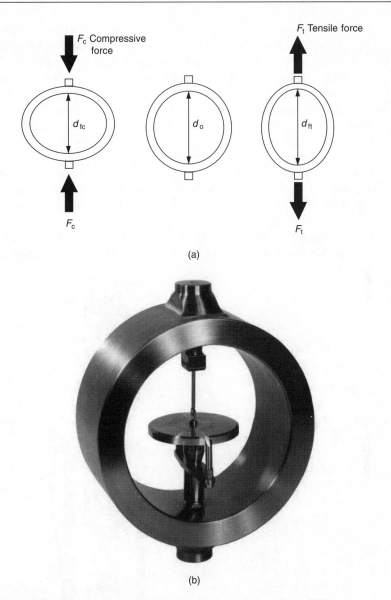

F_c Compressive
force

F_t Tensile force

d_{fc}

d_o

d_{ft}

F_c

F_t

(a)

(b)

Figure 6.3 Proving ring: (a) principle; (b) ring of capacity 1 MN.
(Source: Morehouse Instrument Co., USA)

6.4.4 Universal calibrator

An often-used instrument to transfer forces between a secondary standard
and a working standard or a transducer is the universal calibrator. In this
instrument two devices (e.g. a proving ring and a load cell) can be positioned
and loaded in series with an applied force.

The calibrator consists primarily of three major parts, the stationary frame, the movable frame or yoke and the hydraulic jack (Fig. 6.4). The stationary frame is fixed on the ground. The movable frame can be adjusted to suit the length of the reference standard. In addition the upper platen of the yoke can also be moved up and down to allow for different sized test items. The upper platen is also provided with a ball seat, which allows a hardened steel ball to be positioned between the yoke platen and the reference standard (proving ring) to assure axial loading.

The lower platen of the stationary frame and the lower yoke platen are provided with centre holes. These holes provide a means by which the test items can be attached to the machine with suitable studs or tension members. The machine is designed to minimize friction and non-axial loading, which usually give rise to large errors.

A special hydraulic jack activated by a precision two-speed pump advances the ram quickly until a designated pressure (usually about 5 MPa) is reached

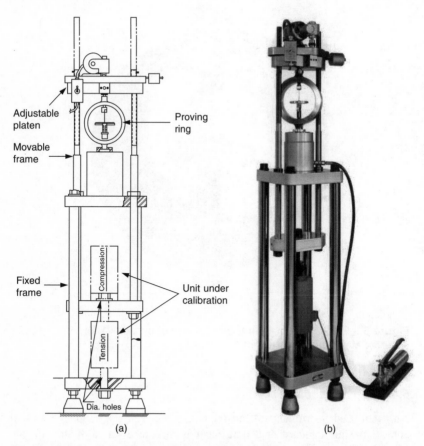

(a)   (b)

Figure 6.4 Universal force calibrator. (a) Schematic; (b) 200 kN capacity calibrator (Source: Morehouse Instrument Company, USA)

and then slowly until the desired load is achieved. The leak rate of the jack is very small and does not affect precise calibrations. The machine can quite easily be adapted to calibrate either in tension or compression mode.

6.5 Force measuring instruments

6.5.1 Characteristics of force measuring devices

The typical output characteristic (response curve) of a force measurement transducer is shown in Fig. 6.5. In this diagram the output of the transducer against the applied force is plotted as the applied force is increased from zero to the rated capacity and returned to zero. A number of significant features of a force measuring transducer or system are illustrated in this diagram.

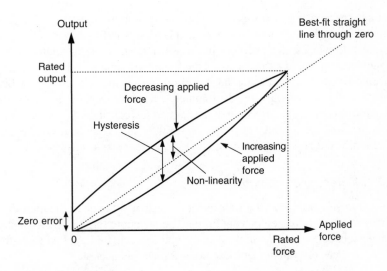

Figure 6.5 Typical response curve of a force measuring transducer

The deviation of the response from a straight line is magnified in this diagram for purposes of clarity. A commonly used method of characterizing a force measuring system is by the use of a best-fit straight line passing through zero.

6.5.1.1 Rated capacity

The *rated capacity* is the maximum force that a force transducer is designed to measure.

6.5.1.2 Non-linearity

Deviations from the best-fit line are referred to as *non-linearity* and usually the largest deviation is given in the specifications of a transducer.

6.5.1.3 Hysteresis

The difference of the output readings between the increasing and decreasing forces at any given force is defined as the *hysteresis*. The largest value of hysteresis is usually at mid range of the system. Sometimes the non-linearity and hysteresis are combined in a single figure. This is usually done by drawing two lines parallel to the best-fit line enclosing the increasing and decreasing curves. The difference of output between the lines is halved and stated as ± *combined error*.

6.5.1.4 Creep and creep recovery

A force measuring system usually takes some time to adjust to a change in applied force. This is known as *creep* and is usually defined as the change of output following a step increase in force from one value to another. Most manufacturers specify the creep as the maximum change of output over a specified time after increasing the force from zero to the rated force, e.g. 0.04 per cent of rated output over 30 minutes. *Creep recovery* is the change of output following a step decrease in the applied force, usually from rated force to zero. Both *creep* and *creep recovery* values are dependent on the time duration of the applied force at the rated capacity or zero respectively.

6.5.1.5 Frequency response

The *frequency response* of a transducer or system is the quantification of its ability to measure forces varying in time, i.e. dynamic forces. The frequency response is defined as the highest sinusoidal frequency of applied force that the transducer or system can measure to a specified accuracy.

6.5.1.6 Fatigue life

If a transducer is used for measurement of fluctuating forces, then its *fatigue life* should be considered. *Fatigue life* is defined as the number of total full cycles of force that may be applied before the measurement uncertainty is altered beyond specified limits.

6.5.1.7 Temperature effects

Both the zero and rated output of a force transducer are affected by a change in the temperature. The *temperature coefficient of the output at zero force* and *temperature coefficient of the sensitivity* are measures of this effect for a given transducer or system.

Other influence quantities such as humidity, pressure, electrical or radio frequency interference may have similar effects to that of temperature and should be taken into account in the design of force measurement systems.

6.5.2 Strain gauge load cell

The most commonly used instrument for force measurement is the electrically operated strain gauge load cell. Strain gauge load cells are available for measurement of tensile, compressive and shear forces. They are also used for measurement of torque. The rated capacity of strain gauge load cells range from 5 N to 50 MN.

6.5.2.1 Principle of operation

A metallic body deforms on the application of a force on it. A tensile force applied on a cylindrical body is shown in Fig. 6.6. There is an increase of the length as well as a small decrease in the diameter. When the applied force is removed the body returns to its original dimensions, provided the elastic limit of the material had not been exceeded. The longitudinal as well as lateral deformations are sensed by strain gauges bonded to the surface of the cylinder.

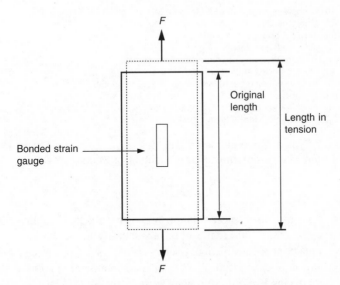

Figure 6.6 Basic principle of the elastic transducer element

6.5.2.2 Elastic element

A number of different shapes of elastic elements are used depending on the range of force to be measured, dimensional limits, final uncertainty and cost of production. A range of commonly used types and their rated capacities is given in Fig. 6.7.

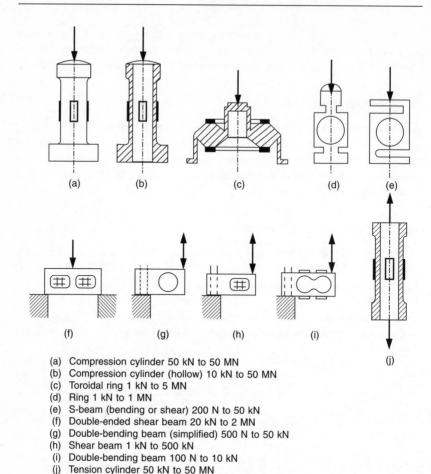

(a) Compression cylinder 50 kN to 50 MN
(b) Compression cylinder (hollow) 10 kN to 50 MN
(c) Toroidal ring 1 kN to 5 MN
(d) Ring 1 kN to 1 MN
(e) S-beam (bending or shear) 200 N to 50 kN
(f) Double-ended shear beam 20 kN to 2 MN
(g) Double-bending beam (simplified) 500 N to 50 kN
(h) Shear beam 1 kN to 500 kN
(i) Double-bending beam 100 N to 10 kN
(j) Tension cylinder 50 kN to 50 MN

Figure 6.7 Typical strain elements and their usual rated capacities. (Source: Guide to the Measurement of Force, Institute of Measurement and Control, UK)

A material having a linear relationship with stress and strain, with low hysteresis, low creep and high level of repeatability in the working range, is used as the material of construction. Usually stainless steel, tool steel, aluminium or beryllium copper is used. A special heat treatment including subzero temperature cycles is required to achieve stability.

6.5.2.3 Resistance strain gauge

In an electrical resistance strain gauge, the change of resistance of an electrical conductor arising from change of its length and cross-section is utilized to detect strain. When a strain gauge is bonded to a metallic substrate, the changes of strain in the substrate will be reflected as a change in resistance of the gauge. This is measured and used to determine the applied force by calibrating the device.

The change in gauge resistance (δR) is related to the change in gauge length (δL) by the gauge factor k:

$$k = \frac{\delta R/R}{\delta L/L} \tag{6.2}$$

where R = gauge resistance and L = gauge length.

For a strain gauge to be useful it should have a relatively high gauge factor, so that small changes in strain give rise to large changes in resistance. Also the gauge factor must be constant over the rated range of applied strains. In addition it must not change significantly over time.

Copper–nickel, nickel–chromium, nickel–chromium–molybdenum and platinum tungsten alloys generally referred to by their trade names are the most common materials used for the manufacture of strain gauges.

A large variety of strain gauges is available for various applications. A strain gauge is usually designed to measure the strain along a clearly defined axis so that it can be properly aligned with the strain field.

The nominal resistance of the strain gauge varies with the type and application. Wire gauges have resistances in the range of 60 Ω to 350 Ω, foil and semiconductor gauges from 120 Ω to 5 kΩ and thin film types around 10 kΩ.

6.5.2.4 Foil strain gauge

The foil strain gauge is the most commonly used type and a number of different designs are shown in Fig. 6.8.

The foil strain gauge is constructed by bonding a sheet of thin rolled metal foil, 2–5 μm thick, on a backing sheet of 10–30 μm thick and photo-etching the measuring grid pattern including the terminal tabs or by cutting the grid from foil using accurate dies. In photo-etching, production techniques similar to those used in the integrated circuit manufacturing industry are used. Accurate and cheap production of complex or very small grid patterns is possible using these processes.

The backing provides electrical insulation between the foil and the elastic element, facilitates handling and presents a readily bondable surface. Typical backing materials are epoxy, polyamide and glass-reinforced epoxy phenolic resins. Some gauges come with an adhesive layer applied to the backing to make it easy for application of the gauge.

In high-precision load cells, the epoxy or epoxy-derived backing material is preferred because of its superior performance, especially creep and low level of moisture absorption compared to polyamide type plastic, although epoxy-based material is difficult to handle due to its brittle nature. Typical specifications of foil gauges are given Table 6.2.

Normally strain gauges are available with automatic compensation that matches the temperature expansion coefficient of one of the three most

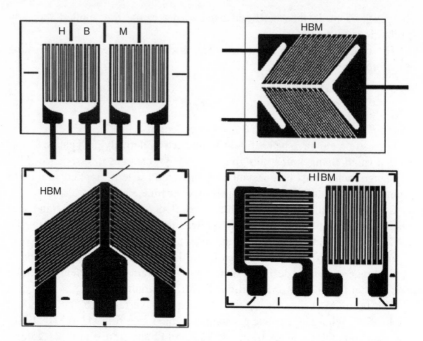

Figure 6.8 Typical metal foil strain gauges. (Source: Hottinger Baldwin Measurements)

Table 6.2 Typical specifications of foil and semi conductor type strain gauges

Characteristic	Typical specifications	
	Foil gauge	Semiconductor gauge
Gauge length, mm	2, 5 or 8	–
Gauge factor	Approximately 2. This is usually quoted individually with the gauge or pack of gauges to two decimal places.	100 to 150 Individually calibrated
Gauge factor temperature coefficient, %/K	±0.015	–
Resistance, Ω	120, 350, 600 and 1000	120
Measurable strain %	2 to 4	up to 5
Fatigue life, strain reversals	up to 10^7 at 0.1% strain	up to 10^6
Temperature range, °C	−30 to +180	–
Temperature compensation: General purpose steels, per K Stainless steels, per °C Aluminium, per °C	11×10^{-6} 17×10^{-6} 23×10^{-6}	

commonly used materials namely general purpose steel, stainless steel and aluminium. Some manufacturers also supply gauges compensated for use on titanium, magnesium alloys and plastic materials. When a temperature compensated gauge is bonded to a material for which the gauge has been matched, the apparent strain due to temperature variations on the gauge can be held down to less than 1.5 micro strain per °C over a temperature range from –20°C to +150°C.

6.5.2.5 Semiconductor strain gauge

Semiconductor strain gauges consist of a strip conductor made from a single crystal of P- or N-type silicon. Due to the high piezo-resistive effect of these materials, their electrical conductivity is highly dependent on the applied strain. The gauge factor of a semiconductor strain gauge is typically 100–150 compared to typically 2–4 of that of a wire or foil strain gauge. The output from semiconductor gauges is non-linear with strain. However, their fatigue life is extremely long and they exhibit only minimal creep or hysteresis.

Due to the relatively high temperature coefficient of these gauges, careful matching of the gauges is required on any given load cell. P-type gauges exhibit positive gauge factors and N-type gauges have negative gauge factors. By combining a P-type with an N-type a temperature compensated pair can be obtained. Usually such pairs are selected by computer matching during manufacture, but compensation circuitry may still be required on the completed transducer. This type of gauge is widely used on small force transducers, accelerometers and pressure sensors. Typical specifications are given in Table 6.2.

6.5.2.6 Thin film strain gauge

Thin films of metals or alloys are deposited on the elastic element using radio frequency sputtering or thermal evaporation techniques for fabrication of thin film strain gauges. The gauge is insulated from the substrate by deposition of a layer of insulating material such as alumina. Several stages of evaporation and sputtering may be used resulting in several layers of material. A number of thin film strain gauge force transducers are available covering a range of 0.1 N to 100 N in the form of single- or double-bending beam configuration.

6.5.2.7 Wire strain gauge

The wire strain gauge made from a wide range of materials is used extensively for high temperature transducers and stress analysis. The wire is typically 20–30 μm in diameter and may be bonded to the substrate using ceramic materials. The 'free' form where the wire is looped around insulated pins mounted on the elastic member is less commonly used. The wire strain gauge is the original type of resistance strain gauge, though now widely replaced by cheaper foil or thin film types

6.5.2.8 Instrumentation

The resistance change of a strain gauge is detected by incorporating it in a Wheatstone bridge configuration as shown in Fig. 6.9. To maximize the response of the load cell one or more strain gauges aligned to respond to the longitudinal strain and another set aligned with the transverse strain are connected in the arms of the bridge. This configuration also minimizes the effects of influence quantities such as temperature that act equally on all the gauges. The resistance change is detected by measuring the differential voltage across the bridge.

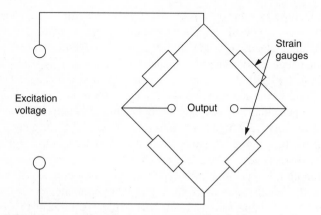

Figure 6.9 Basic arrangement of four strain gauges

The voltage ouput from the bridge when excited by an input voltage is linearly related to the resistance changes of the strain gauges. The output voltage is proportional to the product of the strain and the excitation voltage. The output of a bridge is usually rated to 2 mV/V (2 millivolts output per 2 volts applied), but this can range from 1 mV/V to 4 mV/V.

To realize the full capability of the strain gauge load cell, several correction and compensation components are needed. A circuit diagram incorporating these components as used in a typical commercial load cell is given in Fig. 6.10.

The bridge is usually supplied with a direct current (DC) voltage. The output is amplified using an instrumentation amplifier. This method has wide frequency bandwidth, high stability and relatively low cost. Alternatively the bridge may be excited by an AC voltage, having a sine, square or other waveform. In this case the output is processed through an AC amplifier, a demodulator, filter and DC amplifier. An AC excitation system has a higher immunity to thermo-electric effects in the transducer and thermal effects in the instrumentation. They also have high noise rejection, good zero force output stability and ease of achieving isolation between the signal output and the load cell. However, these systems tend to be costly due to the relatively complex measuring chain.

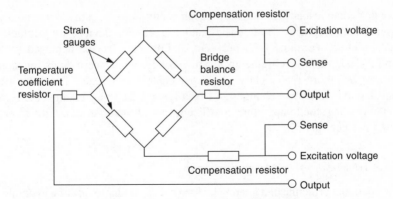

Figure 6.10 Typical commercial load cell circuit

6.5.2.9 Hydraulic load cell

In a hydraulic load cell (Fig. 6.11) the load cell cavity is filled with fluid (usually oil) and given a pre-load pressure. The application of a force to the loading member increases the fluid pressure, which is measured by a pressure transducer.

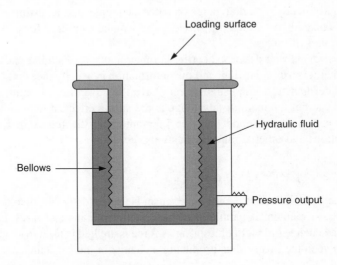

Figure 6.11 Typical hydraulic load cell

Hydraulic load cells are inherently very stiff, deflecting only about 0.05 mm under full force conditions. Capacities of up to 5 MN are available, although most devices are in the range of 500 N to 200 kN. The pressure gauge can be mounted several metres away from the load cell by using a special fluid-filled hose. In systems where more than one load cell is used a specially designed totalizer unit is employed. No external power is needed to operate hydraulic load cells and they are inherently suitable for use in potentially explosive atmospheres. Both tension and compression devices are available. Uncertainties of around 0.25 per cent of full scale can be achieved with careful design and favourable application conditions. Uncertainties for total systems are more realistically 0.5–1 per cent of full scale. The cells are sensitive to temperature changes and usually have facilities to adjust the zero output reading, the temperature coefficients are of the order of 0.02 per cent to 0.1 per cent per °C.

6.5.2.10 Pneumatic load cell

The operating principle of a pneumatic load cell is similar to that of a hydraulic load cell. The force applied to one side of a piston or a diaphragm of flexible material is balanced by pneumatic pressure applied on the other side. The counteracting pressure is proportional to the force and is displayed on a pressure dial.

6.5.2.11 Elastic devices

The 'load column', a metal cylinder subjected to a force along its axis, is a simple elastic device used for measuring forces. The length of the cylinder is measured directly by a dial gauge or other technique, and an estimate of the force is made by interpolating between the lengths measured for previously applied known forces.

The proving ring described earlier is functionally very similar except that the element is a circular ring, and the deformation is usually measured across the inside diameter. These transducers have the advantage of being simple and robust, but the main disadvantage is the strong effect of temperature on the output. Such methods find use in monitoring the forces in building foundations and other similar applications.

6.5.2.12 Capacitive load cell

In capacitive load cells a capacitance sensor is used to detect the displacement of an elastic element. A parallel plate capacitor is used in most cases. In some cases the change of length of a spring as force is applied is used to change the gap between the plates, thus producing a change in the capacitance.

6.5.2.13 Optical strain gauge

In an optical strain gauge, the change in length of an optical fibre is utilized to detect strain.

The deformation of the elastic force-bearing member with the optical strain gauge bonded to it will result in length changes in the optical fibres. If two optical strain gauges experiencing different strain levels are fed with monochromatic light then the phase difference between the two beams emerging from the gauges, in number of half wavelengths, is a measure of the applied force. The advantage of an optical strain gauge is that they are immune to interference by electric and electromagnetic fields.

6.5.2.14 Magnetic transducer

The best known magnetic type load cell transducer is the *Pressductor* cell developed by ASEA of Sweden. In this transducer the change of permeability occurring in a magnetic core due to an applied force is utilized. The principle of the Pressductor is shown in Fig. 6.12.

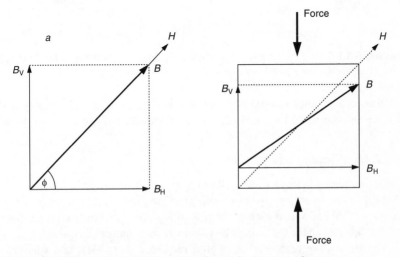

Figure 6.12 Magnetic principle of the Pressductor transducer

A square sheet of transformer iron is magnetized along one diagonal. The material being magnetically isotropic the magnetic flux density vector ($\mathbf{B}$) is parallel to the magnetic field vector ($\mathbf{H}$) and the horizontal and vertical components $B_V = B_H$. When vertical forces are applied anisotropy caused by magneto-elastic effects decreases the permeability in the direction of the forces and $B_V \leq B_H$.

The Pressductor transducer consists of a laminated iron core with two perpendicular windings as shown in Fig. 6.13. An alternating current through the primary winding sets up an alternating magnetic field in the core. However, under no load condition, no voltage is induced in the perpendicular secondary winding. When a force is applied to the core the change in permeability causes the magnetic flux lines to change the angle (φ) generating a voltage in the secondary winding. The induced voltage is directly proportional to the applied force.

The calibration curve of the Pressductor is principally S shaped and linearization is needed to obtain high accuracy. Pressductor load cells having

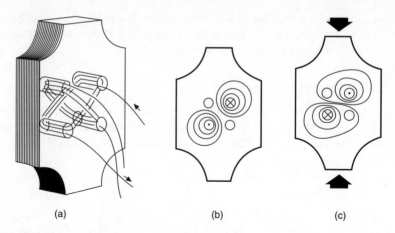

(a)   (b)   (c)

Figure 6.13 The Pressductor transducer. (Source: Electronic Weighing by Ellis Norden, Butterworth-Heinemann)

relative uncertainties of 0.05 per cent of full scale up to 20 tonne-force and 0.1 per cent up to 160 tonne-force are commercially available.

6.5.2.15 *Vibrating strings transducer*

The principle of operation of a vibrating strings transducer is illustrated in Fig. 6.14. The vibrating strings or wires (S) are placed in the air gap of two permanent magnets, and each of them is connected to an electronic oscillator circuit, which causes the strings to vibrate at their natural frequency (f_0).

The strings are preloaded through a reference mass (M), and when the unknown force (F) is applied to the load connection through a string at a

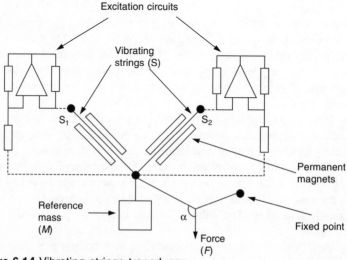

Figure 6.14 Vibrating strings transducer

certain angle (α), the left string (S_1) will be exposed to an increased tensional force, which increases the natural frequency (f_1) of that string, whereas the other string (S_2) will lower its natural frequency (f_2), due to the decrease in the tensional force.

The difference between the two frequencies (δf) = ($f_1 - f_2$) is thus proportional to the applied force (F). The strings are connected to an electronic circuit, which converts the frequency difference (δf) to a pulse train. A direct force reading is achieved by sampling the pulse train in a pulse counter during a predetermined time.

As the vibrating strings transducer compares the unknown force with a known reference force (M), it is independent of the earth's gravity. It has a high linearity and accuracy, and is reported to have extreme long-term stability. This transducer is used in laboratory scales and other scales for smaller weights. Industrial applications include platform and belt conveyor scales.

6.5.2.16 Piezoelectric transducer

When certain types of crystalline materials are subjected to a force, an electric charge proportional to the rate of change of the force is generated on the crystal surface. A charge amplifier is used to integrate the electric charges to give a signal that is proportional to the applied force.

Early piezoelectric transducers used naturally grown quartz but today mostly artificial quartz is used. These devices are often known as *quartz force transducers*. A piezoelectric crystal being an active sensor requires no power supply. Also the deformation of the crystal being very small gives them a good high frequency response.

When packaged as a load washer and compressed under a force of 10 kN a typical piezoelectric transducer deflects only 0.001 mm. The high frequency response (up to 100 kHz) enabled by this stiffness and the other inherent qualities of the piezoelectric effect makes piezoelectric crystal sensors very suitable for dynamic measurements.

Piezoelectric sensors operate with small electric charges and require high impedance cable for the electrical interface. It is important to use the matched cabling supplied with a transducer. A small leakage of charge known as drift is inherent in the charge amplifier. Piezoelectric force transducers are ideally suited for dynamic measurements. Extremely fast events such as shock waves in solids, or impact printer and punch press forces can be measured with these devices when otherwise such measurements might not be achievable. However, they are not very suitable for static measurements. When measurements are taken over a period of minutes or even hours they are said to take 'quasi-static' measurements.

Piezoelectric crystal sensors are suitable for measurements in laboratories as well as in industrial settings. The measuring range is very wide and the transducers survive high overload (typically >100 per cent of full-scale output). The sensors' small dimensions, large measuring range and rugged packaging make them very easy to use. They can operate over a wide temperature range and survive temperatures of up to 350°C.

6.5.2.17 Linear variable differential transducer (LVDT)

The linear variable differential transducer (LVDT) is essentially a transformer that provides an alternating current (AC) output voltage as a function of the displacement of a movable magnetic core. An LVDT is sometimes used within a load cell to measure the displacement of an elastic element instead of using strain gauges. The lack of friction and the low mass of the core result in high resolution and low hysteresis, making this device ideal for dynamic measurement applications.

6.6 Calibration of force standards and test instruments

6.6.1 General considerations

The most important general concepts relating to calibration of force standards and measuring instruments are discussed in this section. The basic principles of calibration are similar for force standards as well as test instruments as both types rely on similar principles of operation discussed earlier in the chapter. The significant differences are in respect of uncertainty, reference standard against which comparison is done, recalibration interval and reference conditions, particularly ambient temperature.

6.6.1.1 Reference standard

The hierarchy of measurement standards used for calibration of force measuring instruments is given in Fig. 6.15.

This diagram gives the next higher level standard that may be used for the calibration of a given standard or instrument. Primary standard is the dead weight force standard. Secondary standards are those calibrated against the primary standard, namely lever or hydraulic force standard machines, proving rings, load cells and other secondary standard force transducers. Proving rings and load cells are also used as force transfer standards. Working standards are proving rings and load cells used for the calibration of measuring instruments.

6.6.1.2 Test uncertainty ratio (TUR)

The test uncertainty ratio required for a particular standard or instrument is an important criterion to consider before a calibration is performed. The uncertainty assigned to a primary, secondary or working standard is usually quoted at 95 per cent confidence level or coverage factor $k = 2$. The combined uncertainty

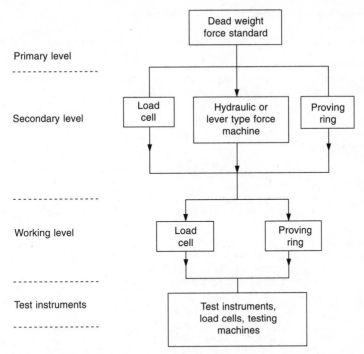

Figure 6.15 Hierarchy of force measurement standards

of the force calibration system including the higher level standard used should be at least three times smaller than the required uncertainty of the standard or instrument under calibration.

In the case of an instrument such as a load cell or an in-built force measurement system, the required uncertainty is determined by the user, taking into consideration the requirements defined by the process (in process control) and regulatory standards (weights and measures, health and safety). Calibration uncertainty should be defined to meet these needs. Typical uncertainties of primary and secondary force standards are indicated in Table 6.3

6.6.1.3 Reference conditions

As all force transducers are temperature dependent, calibrations of force standards are done in a laboratory with controlled temperature and humidity conditions. The reference temperature of calibration of force measurement standards is either 23°C or 25°C.

Whenever possible, force measurement instruments should also be calibrated in a controlled temperature environment. Frequently this is not possible as the test instrument is located in a factory or a site where the ambient temperature is not controlled. In such cases the calibration could be done at the prevailing

Table 6.3 Typical uncertainties of force standards

Type of standard	Principle of operation	Relative uncertainty as % of full scale*	Level of standard
Dead weight standard	Force is generated by suspending a known mass in the earth's gravitational field.	±0.002	Primary
Hydraulic amplification machine	A set of dead weights is used to generate pressure in a small piston cylinder assembly. This pressure is applied to a larger piston cylinder assembly amplifying the force.	±0.04	Secondary
Lever amplification machine	A set of levers is used to amplify the force generated by a small dead weight machine.	±0.04	Secondary
Proving ring	The deformation of an elastic metallic ring on application of force is detected using a dial micrometer or other device.	±0.01	Secondary
Load cell	The deformation of an elastic element on application of force is detected using electrical strain gauges or other device.	±0.05	Secondary

*The uncertainty is at a confidence level of 95%.
(Reproduced from *Guide to Measurement of Force* with permission of the Institute of Measurement and Control, UK)

conditions. However, the temperature of the sensing elements (both test item and standard) should be measured and temperature corrections applied.

6.6.1.4 Range of calibration

The range of calibration should be carefully considered. A force transducer is normally calibrated throughout its total range in tension or compression at least at ten steps. To detect hysteresis effects, increasing force as well as decreasing force is used. Usually calibration over the range 20 per cent to 100 per cent of rated output is more critical.

6.6.1.5 Scope of calibration

A force measurement system generally consists of a force transducer, indicating device, associated cabling and power supplies. Calibration of the entire system is the best solution. However, this is not always practicable, in which event the transducer and instrumentation are calibrated separately to traceable standards.

6.6.1.6 In situ *or laboratory calibration*

Secondary and working level force standards should always be calibrated in a laboratory having temperature and humidity control. In industrial situations, whether to calibrate a force measurement system in a laboratory or *in situ* is dependent on a number of factors. *In situ* calibration is very often necessitated by reasons of cost, to avoid disturbing the instrument or to calibrate exactly under conditions of use.

6.6.1.7 Recalibration interval

The interval between calibrations of a force standard or instrument is dependent on the type of transducer, uncertainty specification and frequency of use. The manufacturer's specification for the device, particularly the specification for the long-term stability, is a good starting point for the determination of the recalibration interval. In most calibration laboratories secondary standard proving rings are given a three-year recalibration interval. Resistance strain gauge load cells, however, require shorter intervals due to their inherent drift characteristics. Generally these are calibrated annually or even at shorter intervals. It is possible to determine the approximate recalibration interval after several calibrations, if *as found* calibration data are recorded in the calibration report. If the calibration values change significantly at each calibration, then the interval may be too long. On the other hand if there is no appreciable change in the calibration values, the interval may be too short. Specific recommendations in respect of working standards are given in the next section.

6.6.2 Calibration of working standard force proving devices

6.6.2.1 Documentary standards

There are a number of international and national standards applicable to the calibration of working standard force proving devices (e.g proving rings and load cells) used for calibration of material testing machines and other force measurement systems. Two widely used standards are:

(a) The international standard ISO 376.
(b) American Society for Testing and Materials standard ASTM E 74.

6.6.2.2 Reference standard

Working standard devices (proving rings and load cells) are calibrated against a secondary standard proving ring, load cell or hydraulic or lever type force standard depending on their rated capacity. A test uncertainty ratio of at least 1:3 should be maintained.

6.6.2.3 Reference temperature

The calibration is conducted in an environment with an ambient temperature of 18°C to 28°C stable to ±1°C. Most laboratories use the temperature of 23°C ± 1°C. Sufficient time is allowed for the device under calibration to achieve temperature equalization with the ambient.

6.6.2.4 Preliminary tests

It is common practice to carry out a few preliminary tests before undertaking the calibration of a proving device. The most common tests are overload, the method of application of the forces and the effect of a variation in supply voltage on the output of the device.

Overload test

The instrument is subjected to an overload within the range of 8 per cent to 12 per cent of the maximum load four times in succession. The overloading is maintained for a duration of 1 minute to 1.5 minutes.

Application of forces

The ability of the attachment system of the force proving instrument to apply the forces axially when the instrument is used in a tensile force application is tested. Also when the instrument is used in the compression mode, it is ensured that the deflection of the force proving instrument is not significantly affected by the characteristics of the bearing pads, namely its hardness and curvature of the bearing surfaces.

Variable voltage test

A change of the supply voltage by ±10 per cent should not produce any significant effect on the output of the force proving device.

6.6.2.5 Number of test loads

At least 30 load applications are required for a calibration and of these at least ten must be at different loads. That is the load is applied at ten different test points, three times, over the full range of the device giving rise to a total of 30 applications. Usually the instrument under calibration is rotated symmetrically on its axis to positions uniformly distributed over a 360° circle at intervals of 120° or 180°.

6.6.2.6 Preload

The preloading is done to establish the hysteresis pattern and is particularly required if the device has been out of use for some time or if the mode of loading is changed say from compression to tension in the case of dual mode devices. The device is subjected to the maximum force three times before the calibration run is started. The preload is maintained for a period of 1 to 1.5 minutes.

6.6.2.7 Load increments

After preloading, the calibration loads are applied starting from zero load and increasing to the highest load. Whether to return to zero load after each calibration load is decided based on the stability of the zero load reading and the presence of noticeable creep under load. However, to sample the device behaviour adequately a return to zero load should be made after application of not more than five consecutive calibration loads. The loading sequence is repeated a number of times (usually two or three times). Before each repeat the position of the device in the calibration machine is changed. A compression device is rotated on its axis by one-third or one-half turn (120° or 180°), keeping the same load axis while in a tensile calibration coupling rods are rotated one-third or one-half turn taking care to shift and realign any flexible connectors.

6.6.2.8 Calibration equation

The deflection of a proving ring or the reading indicated by a load cell for each calibration load is calculated from the following equation:

$$d_L = R_L - (R_{01} + R_{02})/2 \qquad (6.3)$$

where:
d_L = zero corrected deflection or reading of the device for calibration load L
R_L = measured deflection or reading at load L
R_{01} = zero reading before application of load L
R_{02} = zero reading after application of load L.

A polynomial second-degree equation as given below is generally fitted to the load and deflection values obtained from the calibration:

$$d_L = A + BL + CL^2 \qquad (6.4)$$

Other forms and methods of fitting, including higher degree polynomials particularly for high resolution devices are also used.

6.6.2.9 Uncertainty

The uncertainty (U) is calculated from the following equation:

$$U = 2.4 \times s \tag{6.5}$$

where s is the standard deviation of the residuals, i.e. differences between the measured deflections (d_L) and their corresponding values obtained from the calibration fit equation (d_{FL}). The residuals (r_L) are calculated from the equation:

$$r_L = d_L - d_{FL} \tag{6.6}$$

and the standard deviation of n residuals from the equation:

$$s = \sqrt{\frac{\sum_{L=1}^{n} r_L^2}{(n - m)}} \tag{6.7}$$

In this equation m is the number of coefficients in the fitted curve (equation (6.4)).

According to ASTM E74 the factor 2.4 has been determined empirically from an analysis of a large number of force measuring device calibrations and contains approximately 99 per cent of the residuals for least square fits.

An example of data analysis of the calibration of a working standard proving ring is given in Table 6.4.

Table 6.4 Calibration of a working standard proving ring – data analysis

Applied force	Deflection	Temperature	Temperature corrected deflection	Deflection from equation of fit	Residual	Square of residuals	Force per deflection
kN	div.	°C	div.	div.	div.	div²	N/div.
20	28.4	23.6	28.4	28.88	0.48	0.23	0.70
40	74.3	23.2	74.3	73.68	−0.62	0.38	0.54
60	135.2	23.7	135.2	134.48	−0.72	0.52	0.44
80	211.1	23.4	211.1	211.28	0.18	0.03	0.38
100	304.1	23.6	304.1	304.08	−0.02	0.00	0.33
120	414.1	23.2	414.1	412.88	−1.22	1.49	0.29
140	537.9	23.8	537.8	537.68	−0.22	0.05	0.26
160	680.1	23.6	680.0	678.48	−1.62	2.62	0.24
180	836.6	23.7	836.4	835.28	−1.32	1.74	0.22
200	1010.0	23.5	1009.9	1008.08	−1.92	3.69	0.20

Sum of squares of residuals = 10.46 div².

Average force per deflection = 0.36 N/div.

Standard deviation = 2.93 div.

Standard deviation in newtons = 1.05 N

Uncertainty as per cent of capacity = 0.53 per cent

Equation of fit: deflection $(d_L) = 0.08 + 1.04 \times L + 0.04 \times L^2$ (6.8)

The lower loading limit for use as a class AA (see section 6.6.2.11) device is $2000 \times 1.05 = 2100$ N $= 2.1$ kN.

6.6.2.10 Temperature corrections

All mechanical force measuring instruments require temperature corrections when used at a temperature other than the temperature of calibration. The correction is effected using the following equation:

$$d_{Lt} = d_{L0} [1 + k(t - t_0)] \tag{6.9}$$

where:

d_{Lt} = deflection for load L corrected to temperature t
d_{L0} = deflection for load L at calibration temperature t_0
 k = temperature coefficient of the instrument.

For instruments having force transducers made of steel with not more than 7 per cent alloying elements, the value $k = 0.000\ 27/°C$ may be used. In the case of transducers made of material other than steel or having electrical outputs, the temperature coefficient determined experimentally and/or supplied by the manufacturer should be used.

The force corresponding to the corrected deflection at the calibration temperature is obtained from the calibration equation. An example is given below:

Example
Material of the transducer: steel with less than 7 per cent alloying elements
Temperature of the force proving instrument: 25°C
Temperature of calibration: 23°C
Observed deflection at 25°C: 751.4 divisions
Using equation (6.9) the deflection corrected to 23°C (d_{L0}) is obtained:

$$d_{L0} = \frac{751.4}{[1 + (25 - 23) \times 0.000\ 27]} = 751.0$$

Using the calibration equation (6.8) the corresponding force is obtained:

$$751.0 = 0.08 + 1.04 \times L + 0.04 \times L^2$$

The solution of this quadratic equation gives the value of the force as 179.4 kN.

6.6.2.11 Classes of instruments

ASTM E 74 defines two classes of instruments, *Class AA* and *Class A*. Class AA is specified for devices used as secondary standards. The uncertainty of these instruments as determined above should not exceed 0.05 per cent of load. The lower load limit of the instrument is 2000 times the uncertainty in force units, e.g. if an instrument has an uncertainty of 20 N as calculated from equation (6.5), then its lower load limit for use as a Class AA device is 20 × 2000 = 24 000 N. On this basis, the uncertainty of the device for loads greater than 24 000 N would be less than 0.05 per cent.

Class A devices should have uncertainties less than 0.25 per cent. The lower load limit of these instruments is given by 400 times the uncertainty

Table 6.5 Classes of force proving instruments

| Class | Relative error of the force proving instrument % | | | | | Uncertainty of applied calibration force % |
	of reproducibility	of repeatability	of interpolation	of zero	of reversibility	
00	0.05	0.025	±0.025	±0.012	0.07	±0.01
0.5	0.10	0.05	±0.05	±0.025	0.15	±0.02
1	0.20	0.10	±0.10	±0.050	0.30	±0.05
2	0.40	0.20	±0.20	±0.10	0.50	±0.10

(Reproduced from ISO 376–1999 with permission of the International Organization for Standardization)

calculated from equation (6.5). Thus for the instrument having 20 N uncertainty the Class A lower load limit is $20 \times 400 = 8000$ N.

The international standard ISO 376 classifies the instruments into four classes, 00, 0.5, 1 and 2 based on five criteria. This classification is given in Table 6.5.

6.6.2.12 Limited load devices

Elastic rings or other devices with dial indicators for sensing deflection are classified as limited load devices as large localized non-linearities are introduced by their indicator gearing. These devices are to be used only at the calibrated loads and interpolation of values or use of curve fitting techniques are not recommended.

6.6.2.13 Recalibration interval

ASTM E 74 recommends an interval of two years for mechanical force measuring instruments such as proving rings, amsler boxes, rings or loops with dial indicators or optical scales. Electrical force measuring instruments such as strain gauged load cells, rings or loops with differential transformers, variable reluctance sensors and piezoelectric load cells are to be recalibrated one year after their first calibration and thereafter at intervals not exceeding two years provided the changes of deflection between the most recent calibration and those from the previous calibration do not exceed 0.1 per cent of the full load deflection.

An instrument should be recalibrated after repairs or if it has been subjected to an overload higher than the test overload.

6.6.3 Verification of tensile and compressive testing machines

6.6.3.1 Documentary standards

A number of standards give procedures for verification of material testing machines and other force measurement systems. Two widely used standards are:

(a) The international standard ISO 7500-1.
(b) American Society for Testing and Materials standard ASTM E 4.

6.6.3.2 Reference standard

Load cells or proving rings are the commonest devices used for verification of tensile or compressive testing machines. In some instances if the maximum capacity of the testing machine is relatively low, it is possible to use weights of known value and uncertainty.

The force proving instrument should comply with the requirements of ISO 376 and should be equal to or better than the class for which the testing machine is to be calibrated. If dead weights are used then the relative uncertainty of the force generated by these weights should be less than 0.1 per cent.

6.6.3.3 Temperature equalization

A sufficient period of time is allowed for the force proving device to attain temperature equilibrium with the ambient. The temperature of the force proving device should remain stable to within $\pm 2°C$ during each calibration run. It is good practice to attach a thermocouple or a liquid-in-glass thermometer to the force proving device to measure its temperature and effect a temperature correction.

6.6.3.4 Conditioning of the testing machine

The testing machine is conditioned by applying the maximum force three times with the force proving device placed in position.

6.6.3.5 Application of test forces

Generally three series of measurements with increasing forces are taken. At least five test forces between the 20 per cent and 100 per cent of each range are required. If it is necessary to verify below 20 per cent of the range, then test forces at 10, 5, 2, 1, 0.5, 0.2 and 0.1 per cent of the scale down to and including the lower limit of the calibration are applied. The lower limit of the calibration is determined as follows:

Class	Lower limit
0.5	400 × resolution
1	200 × resolution
2	100 × resolution
3	67 × resolution

An elastic force proving device can be used in two different ways. The load is increased until the testing machine readout reaches a nominal graduation. The reading of the elastic proving device is recorded. Alternatively the force may be increased until a preset value of the elastic proving device is reached and the readout of the testing machine is recorded.

The indicator reading is set to zero for both the test machine and the force proving device before each series of measurements. The zero readings are taken again 30 s after the removal of the force.

6.6.3.6 Data analysis

The arithmetic mean of the values obtained for each series of measurements is calculated. The relative accuracy and relative repeatability of the force measuring machine is calculated from these data.

Parameters of assessment

The assessment parameters of testing machines are as given in Table 6.6.

Table 6.6 Parameters for assessment of testing machines

Parameter	Definition
Relative accuracy	$q = \dfrac{F_i - \bar{F}}{\bar{F}} \times 100$
Relative repeatability	$b = \dfrac{F_{max.} - F_{min.}}{\bar{F}} \times 100$
Relative reversibility	$v = \dfrac{F_i' - \bar{F}}{\bar{F}} \times 100$
Relative resolution	$a = \dfrac{r}{F} \times 100$
Relative zero error	$f_0 = \dfrac{F_{i0}}{F_N} \times 100$

where:

F = force value indicated by the force proving device with increasing test force
$\bar{F}$ = arithmetic mean of several values of F
F_i = force indicated by the force indicator of the testing machine with increasing test force
F_i' = force indicated by the indicator of the testing machine with decreasing test force
$F_{max.}$ = highest value of F_i for the same discrete force
$F_{min.}$ = lowest value of F_i for the same discrete force
F_{i0} = residual indication of the indicator of the testing machine
F_N = maximum capacity of the measuring range of the testing machine.
r = resolution of the force indicator of the testing machine.

6.6.3.7 Classes of testing machine range

ISO 7500-1 classifies force testing machines into four classes as given in Table 6.7.

Table 6.7 Classes of testing machine range

| Class of machine range | Maximum permissible value, % Relative error of | | | | Relative resolution |
	Accuracy q	Repeatability b	Reversibility v	Zero f_0	a
0.5	±0.5	0.5	±0.75	±0.05	0.25
1	±1.0	1.0	±1.5	±0.1	0.5
2	±2.0	2.0	±3.0	±0.2	1.0
3	±3.0	3.0	±4.5	±0.3	1.5

(Reproduced from ISO 7500-1: 1999 with permission of the International Organization for Standardization)

Example

The verification data of a tensile testing machine of capacity 100 kN is given in Tables 6.8 and 6.9.

Using the definitions given in Table 6.6, the relative resolution of this range is calculated as follows:

$$\text{Relative resolution} = \frac{0.1}{100} \times 100 = 0.1$$

In the range 20 kN–100 kN the testing machine has the following characteristics:

Table 6.8 Calibration of material testing machine – data analysis

| Test machine indication kN | Proving device reading kN | | | Mean reading of proving device kN | Relative accuracy % | Relative repeatabiliy % |
	Run 1	Run 2	Run 3			
20.0	20.4	20.3	20.3	20.3	1.5	0.5
40.0	40.8	40.9	40.6	40.8	2	0.7
60.0	61.1	61.2	61.3	61.2	2	0.3
80.0	81.2	81.3	81.6	81.37	1.7	0.5
100.0	101.4	101.2	101.7	101.43	1.4	0.5

Table 6.9 Relative zero error

Initial zero	0	0	0
Final zero	0.2	0.1	0.2
Relative zero error, %	0.2	0.1	0.2

Relative accuracy – 2 per cent
Relative repeatability – 0.7 per cent
Relative zero error – 0.2 per cent
Relative resolution – 0.1 per cent

The relative reversibility was found to be 1.2 per cent though not recorded in Table 6.8. From these data the 20 kN–100 kN range of the machine can be classified as belonging to Class 2.

6.6.3.7 Reverification interval

The reverification interval of a force testing machine depends on the type of machine, the level of maintenance and the amount of use. A machine should preferably be verified at intervals not exceeding 12 months. A machine that has been relocated or has been subjected to major repairs should be reverified.

Bibliography

International and national standards

1. ISO 376: 1999 (E) International Standard on Metallic Materials – Calibration of force-proving instruments used for the verification of uni-axial testing machines. International Organization for Standardization.
2. BS 1610: 1992 British Standard – Materials testing machines and force verification equipment.
3. BS 1610: Part I. 1992 British Standard – Specification for the grading of the forces applied by materials testing machines when used in the compression mode.
4. BS 5233: 1986 British Standard – Glossary of terms used in metrology (incorporating BS 2643).
5. BS EN 10002 British Standard on tensile testing of metallic materials.
6. BS EN 10002-1: 1992 British Standard – Method of test (at ambient temperature).
7. BS EN 10002-2: 1992 British Standard – Verification of the force measuring system of the tensile testing machine.
8. BS EN 10002-3: 1995 British Standard – Calibration of force proving instruments used for the verification of uni-axial testing machines.
9. ISO 7500: 1999 International Standard on metallic materials – verification of static uni-axial testing machines. International Organization for Standardization.

Introductory reading

1. *Guide to the measurement of force*, Institute of Measurement and Control, UK, 1998.

7

Measurement of temperature

7.1 Introduction

Temperature is one of the most important measurement parameters from an industrial point of view. Large numbers of temperature measuring instruments such as mercury-in-glass thermometers, dial thermometers, thermocouple sensors, resistance temperature devices (RTDs) and temperature transmitters are used in most industrial systems. It is also an important parameter in health services, for monitoring of the environment and safety systems.

7.2 SI units

The SI base unit for temperature is the kelvin, defined as the fraction 1/273.16 of the thermodynamic temperature of the triple point of water. The degree Celsius is also recognized in the SI system and is defined by the relation:

$$t = T - 273.15 \qquad (7.1)$$

where t is a temperature in degrees Celsius and T is the equivalent kelvin temperature.

7.3 Thermodynamic scale

Temperature is the degree of hotness of an object and is governed by the laws of thermodynamics. The temperature scale based on the first and second laws of thermodynamics is known as the thermodynamic temperature scale. The lowest limit of the thermodynamic scale is absolute zero or 0 kelvin (K). Since the scale is linear by definition only one other non-zero reference point is needed to establish its slope. This reference point was originally defined as the freezing point of water (0°C or 273.15 K). In 1960 the reference point was changed to a more precisely reproducible point, namely the triple point of water (0.01°C).

However, measurement of temperature on the thermodynamic scale is hardly suitable for practical thermometry. In practice there are three main reasons:

(a) It is difficult to measure thermodynamic temperatures. Apart from the technical elaboration required it could take days if not weeks to measure a single thermodynamic temperature.
(b) In the determination of temperatures on the thermodynamic scale large systematic errors occur though these have been minimized in recent years.
(c) The resolution or repeatability of thermodynamic thermometers is not as good as that achieved by many other empirical thermometers.

For these reasons all practical temperature measurements are based on an empirical temperature scale defined by the CGPM from time to time.

7.4 Practical temperature scales

The current practical temperature scale is known as the International Temperature Scale of 1990 (ITS-90). The first practical temperature scale known as ITS-27 was defined in 1927. The ITS-27 was revised in 1948 to become the ITS-48, and in 1968 a more comprehensive revision gave rise to the International Practical Temperature Scale of 1968 (IPTS-68). The ITS-90 is an improvement on the IPTS-68.

7.5 International Temperature Scale of 1990 (ITS-90)

The International Temperature Scale of 1990 came into operation on 1 January 1990 as the official international temperature scale. Temperatures on the ITS-90 are defined in terms of equilibrium states of pure substances (defining fixed points), interpolating instruments and equations that relate the measured property to ITS-90 temperature. The defining fixed points in the range $-38.83°C$ to $1084.62°C$ are given in Table 7.1.

The most easily realized fixed point is that of the triple point of water, which has been assigned the value of 273.16 K (0.01°C).

All the fixed points above this temperature are freezing points except for gallium point, which is defined as a melting point.

A freezing point is the temperature at which a substance changes state from a liquid to a solid. Similarly a melting point is the temperature at which a substance changes state from a solid to liquid. During these transitions the temperature remains constant and a freezing or melting plateau is observed.

The temperature at which solid, liquid and gaseous states of a substance coexist in dynamic equilibrium is known as a triple point. Thus ice, water and water vapour coexist in dynamic equilibrium at the triple point of water.

Table 7.1 Defining fixed points of ITS-90

Material	Equilibrium state	Temperature	
		Kelvin (K)	Celsius (°C)
Mercury (Hg)	Triple point	234.3156	− 38.83
Water (H$_2$O)	Triple point	273.16	0.01
Gallium (Ga)	Melting point	302.9146	29.7646
Indium (In)	Freezing point	429.7485	156.5985
Tin (Sn)	Freezing point	505.078	231.928
Zinc (Zn)	Freezing point	692.677	419.527
Aluminium (Al)	Freezing point	933.473	660.323
Silver (Ag)	Freezing point	1234.93	961.78
Gold (Au)	Freezing point	1337.33	1064.18
Copper (Cu)	Freezing point	1357.77	1084.62

7.5.1 Interpolation instruments

Temperatures at intermediate points in the scale are realized using thermometers known as interpolation instruments. Standard platinum resistance thermometers are designated as the interpolation instruments in the range 13.8033 K (the triple point of hydrogen) to 961.78°C (the freezing point of silver). Beyond 961.78°C the scale is defined in terms of a reference black body radiator.

7.6 Industrial thermometers

A number of different types of thermometers are used in industrial temperature measurements. The main types being:

(a) Thermocouple thermometers.
(b) Resistance thermometers.
(c) Liquid-in-glass thermometers.
(d) Bimetallic thermometers.
(e) Radiation thermometers.
(f) Optical pyrometers.

7.6.1 Thermocouple thermometers

A thermocouple thermometer consists of a thermocouple sensing element producing an electromotive force (emf) connected to a device capable of measuring the emf and displaying the result in equivalent temperature units. Such a system is shown in Fig. 7.1.

7.6.1.1 Principle of operation

When a finite conductor is subject to a steady temperature gradient the energy

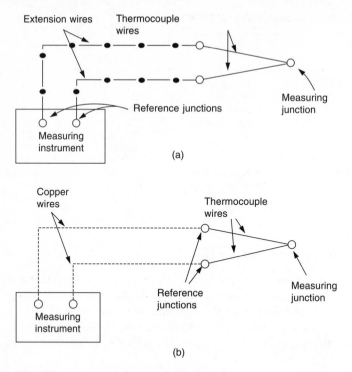

Figure 7.1 Thermocouple thermometer system

levels (Fermi levels) of the electrons which depend on temperature are different. There are more high energy electrons in the hotter region and they move towards the cooler region of the conductor. The resulting separation of charges produces a potential difference at the ends of the conductor. This effect was discovered by Seebeck in 1821 and is known as the Seebeck effect. The potential difference produced per unit temperature difference is known as the Seebeck coefficient and is different for different metals.

If the two junctions of a closed circuit formed by joining two dissimilar metals are maintained at different temperatures, an emf proportional to the difference of the Seebeck coefficients is produced within the circuit (Fig. 7.2). If the temperature of one junction is fixed at some known value, then the temperature of the other junction can be determined by measuring the emf generated in the circuit.

This is the basic principle of thermoelectric thermometry. The junction with the fixed temperature is known as the *reference junction* and is usually kept at 0°C (ice point). The other junction is known as the *measuring junction*.

7.6.1.2 Types and materials

A large number of thermocouple materials have been studied and reported on. The eight types that are used industrially, known as letter designated types,

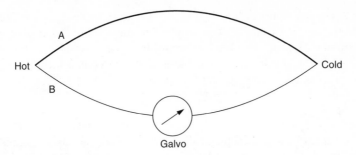

Figure 7.2 Thermoelectric circuit

have been standardized internationally. Table 7.2 gives the standard types and typical uses for them.

The letter designation of the commonly used thermocouple types was introduced by the Instrument Society of America (ISA) and adopted in 1964

Table 7.2 Properties of standardized letter designated thermocouples

Type	Materials	Allowable environment	Maximum operating temperature, °C	Minimum wire diameter, mm
T	Copper/copper–nickel alloy	Oxidizing, reducing inert or vacuum	370	1.63
J	Iron/copper–nickel alloy	Oxidizing, reducing, inert or vacuum	760	3.25
E	Nickel–chromium alloy/copper–nickel alloy	Oxidizing or inert	870	3.25
K	Nickel–chromium alloy/nickel–aluminium alloy	Oxidizing or inert	1260	3.25
N	Nickel–chromium–silicon alloy/nickel–silicon–magnesium alloy	Oxidizing or inert	1260	3.25
R	Platinum–13% rhodium alloy/platinum	Oxidizing or inert	1480	0.51
S	Platinum–10% rhodium alloy/platinum	Oxidizing or inert	1480	0.51
B	Platinum–30% rhodium alloy/platinum–6% rhodium alloy	Oxidizing or inert	1700	0.51

(Reproduced from IEC 60584-Part 2 with the permission of the International Electrotechnical Commission, IEC)

as an American National Standard (C96.1). The letter designations identify the reference tables and may be applied to any thermocouple that has a temperature–electromotive force relationship agreeing within the tolerances specified in the standard. Substantial variations in composition for a given letter type can occur, particularly for types J, K and E.

7.6.1.3 Tolerances

The tolerance of a letter designated thermocouple is defined as the permissible deviation of the emf generated by it at a specific temperature, from the corresponding value of emf given in the reference tables, the reference junction being maintained at 0°C.

Tolerances as given in IEC 60584 are given in Table 7.3. These tolerances apply to new essentially homogeneous thermocouple wire, nominally in the size range 0.25 mm to 3 mm in diameter and used at temperatures not exceeding the maximum temperature given in Table 7.2.

Table 7.3 Tolerances on initial values of electromotive force (emf) vs. temperature of letter designated thermocouples

Types	Temp. range and tolerance value	Class 1	Class 2	Class 3
Type T	Temp. range, °C	−40 to 125	−40 to 133	−67 to +40
	Tolerance value, °C	±0.5	±1	±1
	Temp. range, °C	125 to 350	133 to 350	−200 to −67
	Tolerance value, °C	±0.004 · t	±0.0075 · t	±0.015 · t
Type J	Temp. range, °C	−40 to +375	−40 to +333	−
	Tolerance value, °C	±1.5	±2.5	−
	Temp. range, °C	375 to 750	333 to 750	−
	Tolerance value, °C	±0.004 · t	±0.0075 · t	−
Type E	Temp. range, °C	−40 to +375	−40 to 333	−167 to +40
	Tolerance value, °C	±1.5	±2.5	±2.5
	Temp. range, °C	375 to 800	333 to 900	−200 to −167
	Tolerance value, °C	±0.004 · t	±0.0075 · t	±0.015 · t
Type K, Type N	Temp. range, °C	−40 to +375	−40 to +333	−167 to +40
	Tolerance value, °C	±1.5	±2.5	±2.5
	Temp. range, °C	375 to 1000	333 to 1200	−200 to −167
	Tolerance value, °C	±0.004 · t	±0.0075 · t	±0.015 · t
Type R, Type S	Temp. range, °C	0 to 1100	0 to 600	−
	Tolerance value, °C	±1	±1.5	−
	Temp. range, °C	1100 to 1600	600 to 1600	−
	Tolerance value, °C	± (1+0.003(t−1100))	±0.0025 · t	−
Type B	Temp. range, °C	−	−	600 to 800
	Tolerance value, °C	−	−	±4
	Temp. range, °C	−	600 to 1700	800 to 1700
	Tolerance value, °C	−	±0.0025 · t	±0.005 · t

t = modulus of temperature in °C.
(Reproduced from IEC 60584-Part 2 with the permission of the International Electrotechnical Commission, IEC)

7.6.1.4 Sheathing

As bare thermocouple wire is not suitable for direct use a protective layer, sheathing, or a protective tube (thermowell) should be used. The types of sheathing, protective shields and thermowells available are well described in manufacturers' literature. One of the main advantages of thermocouples is that their rapid response time is degraded by shielding.

One of the most common types in current use, the metal sheathed, ceramic insulated thermocouple cable, provides good electrical and mechanical protection. The sheathing also provides some protection against contamination. However, the metal and insulation could affect the thermocouple at high temperatures due to metal migration. An alumina sheathing, which can be operated up to 1600°C, is available for use with type R and type S thermocouples.

The sheathing and end capping processes may give rise to extra strains in the wire and may not be entirely removed by annealing (prolonged heating at a uniform temperature).

7.6.1.5 Extension and compensation leads

In Fig. 7.2 the measuring instrument is directly connected between the hot and cold junctions of the thermocouple circuit. However, this is not always practicable as the measuring junction may be at a considerable distance away from the measuring unit (Fig. 7.1), and connecting wires have to be used. The same type of material as in the thermocouple should be used for the connecting wires so that extraneous emfs generated due to dissimilarity of the metals are minimized.

For base metal thermocouples (types E, J, K, T and N) extension wires, of the same material as those of the thermocouple elements in the form of flexible multistranded wire, are used. For the more expensive rare metal thermocouples (types R, S and B) the connecting wires in the form of compensation leads made of cheaper materials (copper–based alloys) can be used. Compensation leads having a thermoelectric behaviour similar to that of the thermocouple leads over a narrow temperature range; usually less than 50°C are commercially available.

7.6.1.6 Reference (cold) junction

For very accurate work, such as calibration of thermocouples, the reference junction is kept in an ice bath. However, in most modern thermocouple thermometers cold junction compensation is incorporated. The cold junction is made to remain at a constant temperature by attaching it to a thick copper block (isothermal block) and its temperature is measured with an electric temperature sensor. A silicon sensor is commonly used for the purpose. A correction to the emf generated between the measuring and cold junctions is calculated using the measured temperature of the cold junction and standard temperature-millivolts tables (or equivalent regression equations). This is usually done by a microprocessor incorporated within the instrument.

7.6.1.7 Measurement of thermocouple output

The output of a thermocouple is mostly in the millivolt range. Therefore reliable measurement requires a voltmeter with high input impedance or a potentiometer. In calibration laboratories the output emf is measured using a digital voltmeter or potentiometer. In industrial situations the output is either fed into a digital temperature display or temperature transmitter.

7.6.2 Resistance thermometers

7.6.2.1 Principle of operation

A change in temperature of a resistive element changes its resistance value. This principle is utilized in a resistance thermometer. In the most common type of resistance thermometers, platinum is used as the material of the resistive element. Copper and nickel resistance thermometers are also used in industrial applications.

7.6.2.2 Platinum resistance thermometers

Platinum resistance thermometers (PRTs) are available in two grades, precision grade and industrial grade.

Precision grade platinum resistance thermometers are used as interpolating instruments in the ITS-90 temperature scale and are known as Standard Platinum Resistance Thermometers (SPRTs). These thermometers are of very good stability and precision and are used as secondary standards in national standards laboratories.

SPRTs are available having nominal resistances of 25 ohms and 100 ohms at 0°C. Some manufacturers have recently introduced a new type of thermometer of 0.25 ohms resistance, having a temperature range of 0–1000°C. These thermometers, which are expensive and fragile, are only suitable for use in primary level calibration facilities. However, more robust stainless steel sheathed SPRTs are also available.

The industrial grade platinum resistance thermometers are of more robust construction and are widely used for measurement of temperature in industrial situations.

7.6.2.3 Pt-100 resistance thermometers

The class of thermometers known as Pt-100 is widely used for industrial temperature measurements. The requirements of these thermometers are given in IEC Publication 60751. These thermometers have a nominal resistance of 100 ohms at 0°C. They come in two main categories, wire wound and film types.

Wire wound type

The best wire wound thermometers conform closely to the construction pattern used in working standard platinum resistance thermometers. They are constructed with an alloy comprising pure platinum alloyed with other platinum group metals to make the temperature coefficient close to 0.003 850, which is the value recommended by the standard IEC 60751.

A bifilar winding is wound around a glass or ceramic bobbin, attached to leads and sealed by a layer of glass. This type is very rugged and can withstand high vibration. However, this type of construction is subject to strain during temperature cycling and also the resistive element is not directly in contact with air.

In another form of construction a fine coil of platinum wire is led through holes in an alumina tube and attached to more robust leads. The coil is either partially supported or totally supported along its length and the leads are sealed in place with glass or ceramics.

In the partially supported form of construction the wire is wound into a very small coil and inserted into axial holes in a high purity alumina rod. A small quantity of glass adhesive is introduced into these holes, which, after firing, firmly secures part of each turn into the alumina. This results in a detector in which the majority of the platinum wire is free to move giving very good long-term stability and ability to withstand considerable vibration levels (up to 30g). Figure 7.3 shows the construction details of a partially supported detector.

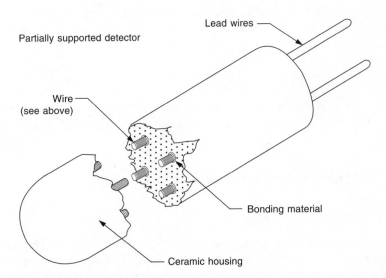

Figure 7.3 Construction details of a partially supported industrial grade platinum resistance detector. (Source: Isothermal Technology Ltd, UK)

Film type

Film type thermometers are constructed using two main techniques. Thick film detectors are made by spreading a glass/platinum paste through a silkscreen mask onto a substrate. Thin film detectors are fabricated by evaporating a metal or alloy onto a substrate, usually alumina in a vacuum chamber.

7.6.2.4 Range

The industrial grade platinum resistance thermometer can be used over the range 0°C to 850°C and –200°C to 0°C. The latter range thermometers have to be sealed to prevent moisture accumulation shorting out the leads.

7.6.2.5 Tolerance

The tolerance of an industrial grade PRT is defined as the maximum allowable deviation expressed in °C from the nominal resistance–temperature relationship. For sensors conforming to IEC 60751 the tolerances are computed from the following equations:

Class	Tolerance, °C
A	$\pm(0.15 + 0.002 \cdot t)$
B	$\pm(0.3 + 0.005 \cdot t)$

In these equations t is the modulus of the sensor temperature in °C. Tolerances worked out from the above equations for sensors of 100 ohms nominal value are given in Table 7.4. For 100 ohm sensors class A tolerances are not specified above 650°C. Also thermometers with only two connecting wires are not categorized in class A.

7.6.2.6 Terminals

PRT sensors are manufactured with two, three or four terminals. The terminal colour coding and sensor identification system given in IEC 60751 is given in Fig. 7.4.

7.6.3 Liquid-in-glass thermometers

7.6.3.1 Principle of operation

Liquid-in-glass thermometers make use of the expansion of a liquid in a glass capillary tube to sense the temperature of the bulb. The temperature is obtained by reading the position of the meniscus from a scale engraved on the capillary stem.

Table 7.4 Tolerances for industrial grade platinum resistance sensors

| Temperature | Tolerance | | | |
| | Class A | | Class B | |
°C	±°C	±Ω	±°C	±Ω
−200	0.55	0.24	1.3	0.56
−100	0.35	0.14	0.8	0.32
0	0.15	0.06	0.3	0.12
100	0.35	0.13	0.8	0.30
200	0.55	0.20	1.3	0.48
300	0.75	0.27	1.8	0.64
400	0.95	0.33	2.3	0.79
500	1.15	0.38	2.8	0.93
600	1.35	0.43	3.3	1.06
650	1.45	0.46	3.6	1.13
700	–	–	3.8	1.17
800	–	–	4.3	1.28
850	–	–	4.6	1.34

(Reproduced from IEC 60751 with the permission of the International Electrotechnical Commission, IEC)

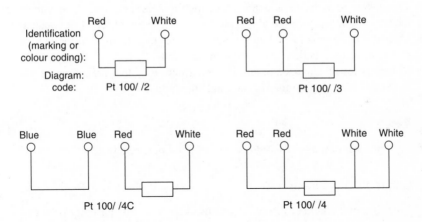

Figure 7.4 Terminal configurations and colour coding of industrial grade platinum resistance sensors. (Source: IEC 60751)

7.6.3.2 Types

Liquid-in-glass thermometers belong to several different types manufactured to conform to a number of national and international standards. The International Organization for Standardization (ISO), British Standards (BS), American Society for Testing and Materials (ASTM), and the Institute of Petroleum (IP) are the most common standard specifications.

Thermometers filled with mercury are the most common type. Other liquids, e.g. alcohols, are sometimes used instead of mercury to extend the range.

Industrial type thermometers are usually encased in metallic enclosures.

7.6.3.3 Range

Mercury-in-glass thermometers can be obtained having a measuring range of –20°C to about 600°C. The finest resolution obtainable is 0.01°C over limited temperature ranges. Spirit filled thermometers can be used down to –200°C.

7.6.3.4 Construction

A liquid-in-glass thermometer consists of a glass bulb filled with the liquid (mercury or spirit) joined to a capillary tube having a uniform bore and sealed at the end. In mercury-in-glass thermometers the space above the mercury is usually filled with nitrogen to minimize column breaks and prevent the mercury from boiling over at higher temperatures. Other essential features of good quality thermometers are as follows.

Glass

The glass used in the manufacture of the thermometer must be able to withstand the maximum temperature and should have a small coefficient of expansion. Usually the glass is identified by one or two coloured lines on the length of the bulb, or by a coded inscription on the stem.

Properties of glasses suitable for construction of thermometers are given in ISO 4795 and a number of national standards.

Bore

The diameter of the bore should be uniform without reductions or constrictions.

Contraction chamber

A contraction chamber (an enlargement of the bore) is normally provided if the thermometer has two scale segments, e.g. one near 0°C and another, say, between 50°C and 100°C.

Expansion volume

An expansion volume may also be incorporated at the top of the bore. This may be a tapered enlargement or simply a suitable extension of the capillary above the highest scale line.

Graduations

Scale lines should be clearly and durably marked, evenly spaced, and of uniform thickness not exceeding one-fifth of the interval between consecutive lines.

Immersion depth

The reading obtained from a liquid in glass thermometer depends to some extent on the mean temperature of the emergent liquid column. Three conditions of immersion are used:

(a) *Total immersion thermometer* – the thermometer is immersed to within a few millimetres of the meniscus when a reading is taken.
(b) *Partial immersion thermometer* – the thermometer is immersed only to a designated mark on the stem.
(c) *Complete immersion thermometer* – requires the entire body of the thermometer to be immersed.

7.6.3.5 Stem corrections

A thermometer should be used vertically and in the condition of immersion given in the calibration certificate (or engraved on the stem). When an immersion depth different to that of calibration has to be used for some reason, a correction is required to take account of the temperature of the emergent liquid column. This is known as the stem correction. The stem correction is worked out using the formula:

$$\delta t = kn(t_1 - t_2) \tag{7.2}$$

where k is the apparent thermal expansion coefficient of the liquid in the type of glass from which the stem is made, t_1 and t_2 are the average temperatures of the emergent columns during calibration and use, respectively, and n is the number of degrees equivalent to the length of the emergent liquid column.

For most mercury-in-glass thermometers $k = 0.000\,16/°C$ and for spirit in glass thermometers $k = 0.001/°C$. The temperature of the emergent column is measured either by the use of a special thermometer with bulb length slightly longer than the emergent column (Faden thermometer) or a short stem thermometer. Some calibration laboratories use thermocouples to measure stem temperatures during calibration. The method of using a Faden thermometer to measure the emergent stem temperature is shown in Fig. 7.5.

In the example shown in Fig. 7.5, the length of the emergent column of the main thermometer is 150 mm and a Faden thermometer of bulb length 160 mm is placed alongside the main thermometer with 10 mm of its bulb in the comparison bath. This ensures that the Faden thermometer bulb simulates the condition of the emergent stem. The average temperature of the emergent liquid column (t_1 or t_2 of equation (7.2)) is calculated as follows:

$$\frac{55.1 \times 160 - 202 \times 10}{160} = 42.5°C$$

7.6.3.6 Secular change

In a glass thermometer the bulb is continuously recovering from strains

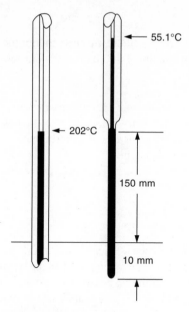

Figure 7.5 Use of a Faden thermometer to measure emergent stem temperature of a partially immersed thermometer

introduced during manufacture, when the glass was heated to 500°C–600°C. This recovery manifests itself as a contraction of the bulb and is known as secular change. In a new thermometer the contraction is relatively rapid and the rate decreases with time. The contraction of the bulb affects the readings significantly as the bulb volume is large in comparison to the stem volume. An ice point check is a useful method of tracking the changes in reading due to this effect. Figure 7.6 shows the effect of secular change on new and well-annealed thermometers. In a well-annealed thermometer, the drift due to secular change is normally less than one scale division per five years.

7.6.3.7 Temporary depression of zero

Glass is basically a liquid but in solid form at ambient temperature (super cooled liquid). Its molecules are relatively free to move, even at ambient temperature. Heating of glass expands it but subsequent cooling does not contract it back to its original volume. The effect occurs every time a thermometer is heated and is called the 'temporary depression of zero'. The temporary depression is normally about 0.1 per cent of the reading or smaller, and lasts for a few days with residual effects detectable for months.

7.6.3.8 Stability and accuracy

The secular change and temporary depression of zero are the main effects contributing to the instability of liquid-in-glass thermometers. However, well-

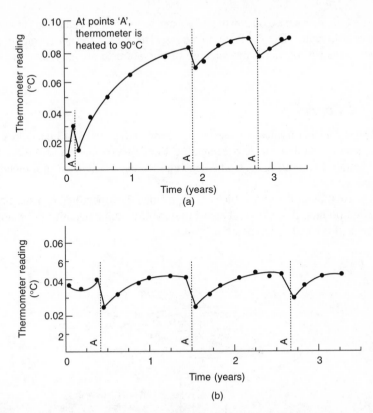

Figure 7.6 Effect of secular change on (a) new and (b) well-annealed thermometers. (Reproduced from Techniques for the Calibration of Liquid-in-Glass Thermometers, Commonwealth Scientific and Industrial Research Organization, Australia)

annealed and calibrated thermometers are usually accurate to about 0.2–0.5 scale divisions. Accuracies of ASTM thermometers are specified in ASTM E1 and those of reference quality thermometers in British Standard 1900. In general a good practical guide for the scale error of a thermometer is its resolution. A thermometer whose scale error is more than one resolution should be discarded.

7.6.4 Bimetallic thermometers

7.6.4.1 Principle of operation

The difference in the thermal expansion of two metals is utilized in the operation of a bimetallic thermometer. The bimetallic sensor consists of a strip of composite material wound in the form of a helix. The composite material consists of dissimilar metals fused together to form a laminate. The difference in thermal expansion of the two metals produces a change in curvature of the strip with changes in temperature. The helical construction of the bimetallic strip translates this change of curvature to a rotary motion of a shaft.

7.6.4.2 Construction

A bimetallic thermometer consists of an indicating or recording device, a sensing element known as a bimetallic thermometer bulb and means for connecting the two, Fig. 7.7. The sensing element is enclosed in a metallic protective case.

The rotation of the bimetallic sensing element is transmitted to a shaft and attached pointer. A scale graduated in temperature units enables the pointer rotation to be read as temperature values.

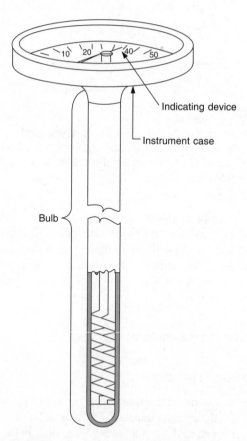

Figure 7.7 Construction of a bimetallic thermometer

7.6.4.3 Types

Basically there are two types of bimetallic thermometers, industrial type and laboratory or test type.

Industrial type

Industrial type thermometers are generally available with 25 mm ($^1/_2$ in) or 75 mm ($^3/_4$ in) standard pipe thread connections. The bulb diameter varies from 3 mm to 10 mm. Bulb lengths from 60 mm to 1.5 m are available.

Laboratory or test type

Laboratory or test type thermometers are of higher accuracy than the industrial type.

Both types are available as straight and angled thermometers. Thermometers are also available with extra pointers that indicate maximum and minimum temperatures. In some thermometers, the stem is filled with silicone oil for providing shock and vibration protection.

7.6.4.4 Range

Bimetallic thermometers are available in the temperature range $-100°C$ to $540°C$. However, they are not suitable for continuous operation above $400°C$.

7.6.4.5 Accuracy

The accuracy of bimetallic thermometers depends on a number of factors, the design environment, immersion, thermal stability of the bimetal element, etc. Generally thermometers having an accuracy of $±1$ per cent of reading are available.

7.6.5 Radiation thermometers

7.6.5.1 Principle of operation

All objects with a temperature above absolute zero emit radiant energy from their surface. As the temperature of the object increases more and more energy is emitted eventually emitting visible energy at around $650°C$. The relationship between wavelength of the emitted radiation and spectral radiance is shown in Fig. 7.8.

Figure 7.8 illustrates two fundamental physical principles:

(a) The amount of thermal energy possessed by an object increases as the temperature of the object increases. The area under the curve represents this.

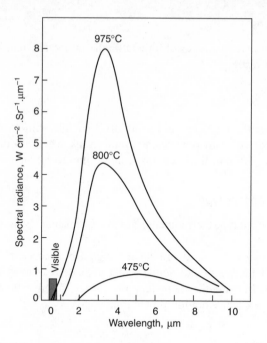

Figure 7.8 Black body radiation characteristics

(b) As the temperature increases, more energy is emitted at progressively shorter wavelengths until, at around 650°C, the radiant energy is in the form of visible light.

In a radiation thermometer these basic principles are utilized to measure temperature. The radiant energy is focused by a suitable lens onto a radiation detector. The electrical signal produced by the detector, which is proportional to the incoming energy, is transmitted to a recording or indicating device.

The types of radiation detectors used today and their characteristics are given in Table 7.5.

The characteristics shown reinforce a basic rule of thumb: short wavelength = high temp; long wavelength = low temperature.

7.6.5.2 Emissivity

Emissivity is a parameter that characterizes the amount of energy an object emits from its surface. It is defined as:

$$\varepsilon_\lambda = \frac{\text{Spectral radiance of object at wavelength } \lambda}{\text{Spectral radiance of black body at wavelength } \lambda}$$

$$\varepsilon_\lambda = \text{Emissivity of object surface at wavelength } \lambda \qquad (7.3)$$

In practice all objects (other than a black body) emit less than 100 per cent of

Table 7.5 Types of radiation detectors and their characteristics

Detector	Wavelength µm	Temperature range °C	Response speed ms
Silicon	0.7 to 1.1	400 to 4000	10
Lead sulphide	2.0 to 2.6	100 to 1400	10
Pyroelectric	3.4 to 14	0 to 1500	100
Thermopile	1 to 14	−150 to + 500	100

(Reproduced from *Measurement and Control* with the permission of the Institute of Measurement and Control, UK)

their energy, so emissivity values are always less than 1.00. The only object that emits all of its thermal energy is a black body. Black body radiators are specialist designs used as the standard for calibrating radiation thermometers.

The accuracy of the values displayed by a radiation thermometer depends on the emissivity of the object being measured. The emissivity–wavelength relationship is non-linear for a majority of objects. The normal trend is for emissivity to be higher at shorter wavelengths than at longer wavelengths. Also as the temperature of an object changes its emissivity can change.

Although this causes problems with measurement accuracy, the expected error can be quantified as a function of wavelength and temperature. Table 7.6 indicates the error in °C, for 1 per cent change in emissivity, related to temperature and wavelength of most commercially available radiation thermometers.

The obvious fact from Table 7.6 is that using the shortest possible wavelength will minimize the effects of emissivity variation on the temperature being measured.

Table 7.6 Change in temperature reading caused by 1 per cent change in emissivity

Temperature °C	Wavelength µm 0.65	0.9	1.6	2.3	3.4	5.0	10.6
100	0.06	0.08	0.15	0.22	0.33	0.49	1.00
500	0.27	0.37	0.68	0.96	1.40	2.10	3.60
1000	0.74	1.00	1.80	2.60	3.70	5.10	7.80
1600	1.60	2.20	4.0	5.50	7.50	9.60	13.00

(Reproduced from *Measurement and Control* with the permission of the Institute of Measurement and Control, UK)

7.6.5.3 Types

A large variety of radiation thermometers are available, with variations in temperature measurement range, detector type, operating wavelength and lens material. It is impossible to summarize all the options available. Table 7.7 shows a broad overview of types available related to their possible uses in industry.

Table 7.7 Types of radiation thermometers and their possible uses in industry

Wavelength µm	Temperature range °C	Detector	Lens	Primary areas of application
0.65	700–3500	Silicon	Crown glass	Molten steel and molten glass
0.7–1.08	550–3000	Silicon	Crown glass	Iron, steel foundries and semiconductor
0.9–1.08	300–2800	Silicon	Crown glass	Silicon wafers and high temperature steel requiring wide ranges
0.91–0.97	400–2000	Silicon	Crown glass	Specifically designed for gallium arsenide temperature in MBE and MOCVD
0.7–1.08 1.08 two colour	700–3500	Silicon	Special lens	High temperature applications in dusty or smoky atmospheres or small targets, e.g. kilns, vacuum furnaces
1.64	250–1100	Germanium	Crown glass	Best choice for non-ferrous metals
2–2.6	80–800	Lead sulphide	Crown glass	Low temperature metals and small targets
3.4	0–800	Indium arsenide	Calcium fluoride	Thin film organic plastics, paints, waxes, oils
3.9	300–1300	Thermopile	Calcium fluoride	Furnace walls of glass melters
4.8–5.3	50–2500	Pyroelectric thermopile	Zinc sulphide	Glass surface temperature for sealing, bending, annealing, tempering and forming
7.9	20–400	Pyroelectric thermopile	Zinc sulphide	Thin films of polyester and fluorocarbons, thin glass and ceramics
8–14	–50–500	Pyroelectric thermopile	Zinc sulphide	General purpose, low temperature, paper, food, textiles.

(Reproduced from *Measurement and Control* with the permission of the Institute of Measurement and Control, UK)

7.6.5.4 Ratio thermometers

A ratio thermometer consists of two detectors working at two different wavelengths. For this reason they are also known as two colour thermometers.

The ratio of the outputs of the two detectors (V_1/V_2) is given approximately by the following equation:

$$\frac{V_1}{V_2} = f(T)\frac{\varepsilon_1}{\varepsilon_2} \times \frac{\lambda_1}{\lambda_2} \qquad (7.4)$$

where T is the target temperature and ε_1 and ε_2 are emmisivities of the target at wavelengths λ_1 and λ_2.

Since the wavelength ratio λ_1/λ_2 is a constant and the emissivity ratio $\varepsilon_1/\varepsilon_2$ is nearly equal to 1 the ratio of the outputs can be written as:

$$\frac{V_1}{V_2} = k \times f(T) \qquad (7.5)$$

which shows that the ratio of the outputs is proportional to the target temperature. The reading of a ratio thermometer is therefore largely unaffected by attenuation of the radiation by dust, smoke or steam, providing the two detectors both receive the same amount of attenuation. For this reason ratio thermometers are an excellent choice for situations where the target is obscured by smoke, steam or dust or the target area of the thermometer is not completely filled, i.e. the target is small or a thin wire.

However, ratio thermometers are also affected by the variation of emissivity of the target. The effect on temperature due to variation of emissivity during a process is something that can only be minimized not eradicated.

7.6.5.5 Portable instruments

Portable radiation thermometers have been available since the early 1970s. The typical features of a portable radiation thermometer are variable focus, and through the lens sighting, with definition of the target area to be measured by a reticle in the viewfinder. This is essential to ensure accurate sighting of the target area. In most types the target temperature is displayed in the viewfinder. Although a variety of models are available two particular model specifications are common. These are given in Table 7.8.

Table 7.8 Typical specifications of portable radiation thermometers

Detector	Wavelength	Temperature range	Field of view
Silicon	0.7 to 1.0 μm	600°C to 3000°C	Distance to target/100
Thermopile	8 to 14 μm	−50°C to +1000°C	Distance to target/40

(Reproduced from *Measurement and Control* with the permission of the Institute of Measurement and Control, UK)

The high temperature instrument is particularly used in the steel, glass, ceramics and metal processing industries, while the lower temperature instrument is used in a wide variety of process industries such as food, plastics and paper.

7.6.5.5 Fixed instruments

A fixed installation radiation thermometer is used to monitor a single position, and feed its output either to a remote temperature display, or to a process control system.

7.6.5.6 Scanning instruments

A scanning radiation thermometer is used to measure the temperature profile of a broad target such as a strip of steel or sheet of glass. A line scanner can be used to measure the temperature of a moving target along a line perpendicular to the direction of travel. If a number of such crosswise profiles are obtained it is possible to generate a two-dimensional thermal map of the object being measured. The outputs of the scanners are either analogue 4 to 20 mA or RS 485 signals, which can be interfaced with plant control systems.

7.7 Calibration of thermometers

The calibration of thermometers of all types is carried out by two main methods, fixed point calibration and comparison calibration.

7.7.1 Fixed point calibration

Fixed point calibration is performed by using ITS-90 fixed points in a fixed point bath or furnace. Carefully performed fixed point calibrations yield very small uncertainties and are usually required only for those thermometers that are used as secondary standards. For most instruments used in industry fixed point calibration is not required.

7.7.2 Comparison calibration

Comparison calibration is performed in a calibration bath in comparison with a secondary or working standard thermometer. Almost all industrial temperature measuring instruments are calibrated by this method.

Due to the importance of these techniques for ISO 9000 certification a brief description of the essential techniques and frequently used calibration equipment are given.

7.7.3 Essential features of calibration

7.7.3.1 Calibration hierarchy

The hierarchy of calibration standards used for calibration of temperature measuring instruments is given in Fig. 7.9.

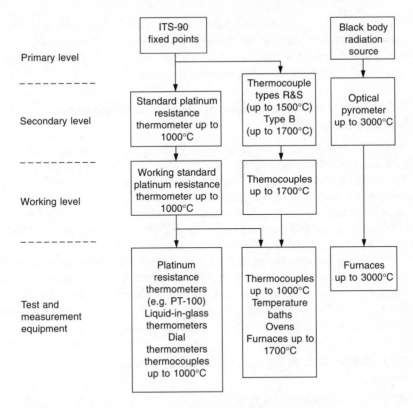

Figure 7.9 Hierarchy of temperature measurement standards

Figure 7.9 gives the next higher level standard that should be used for calibration of a given standard or instrument. Primary standards are those designated by the ITS-90 definition. Secondary standards are those calibrated against the primary standards, namely standard platinum resistance thermometers (SPRTs) up to 1000°C, Type R or S thermocouples up to 1600°C and optical pyrometers up to 3000°C.

Working standards are generally platinum resistance thermometers (up to 1000°C), thermocouples up to 1700°C and reference standard mercury-in-glass thermometers (up to 500°C). However, mercury-in-glass thermometers are rarely used as working standards nowadays due to the availability of low cost very stable platinum resistance thermometers.

In the high temperature area (1600°C to 3000°C) the optical pyrometer is used as the secondary standard.

7.7.3.2 Test uncertainty ratio

The choice of the reference standard and the procedure for calibration is mostly dependent upon the uncertainty of the item under calibration. Generally it is the practice of accredited metrology laboratories to achieve a test uncertainty ratio of at least 1:4, i.e. the uncertainty of the reference standard is at least one-quarter that of the item under calibration. However, in some cases this may not be achieved. In thermometry, absolute uncertainty of the calibrated item is more significant and is generally specified.

7.7.4 Calibration of industrial grade platinum resistance thermometers (PRTs)

A good indication of the validity of calibration of a platinum resistance thermometer (PRT) could be obtained from an ice point check (see section 7.7.6.1). In addition it is good practice to calibrate the sensor periodically in comparison with a working standard PRT over the range of temperature measured. As a general rule PRTs should be calibrated yearly or whenever a significant change in ice point occurs. The availability of portable solid block calibrators has greatly facilitated on-site calibration.

The recent introduction of slim metal clad fixed point cells, which can be used in combination with small, lightweight (bench top or cart mounted) furnaces, has also facilitated the calibration of industrial grade PRTs. These fixed point cells are available for mercury, gallium, indium, tin, zinc and aluminium points with uncertainties ranging from ±0.001°C to ±0.005°C.

7.7.5 Calibration of liquid-in-glass thermometers

7.7.5.1 Recalibration interval

The period of validity of calibration of a liquid-in-glass thermometer is dependent on a number of factors, namely:

(a) The secular change – this is dependent on the age of the thermometer and the degree of annealing.
(b) The uncertainty permissible on the reading.
(c) The pattern of usage of the thermometer, particularly rough handling or wide temperature cycling.

A regular ice point test, at least once in six months, would reveal any changes of calibration occurring due to the secular change of the thermometric glass. If a significant change has occurred it is best to recalibrate the thermometer over the entire working range. If the thermometer has been subjected to rough handling or wide temperature fluctuations the calibration may change and in these cases a recalibration is called for.

After a few calibrations it is possible to determine an optimum recalibration interval by examining the calibration data. If the correction terms are essentially the same, it could be assumed that the thermometer is stable and the recalibration period can be extended. If on the other hand a more or less constant change at all test points is shown, the thermometer is still not stable and frequent ice point checks and yearly recalibration would be required.

7.7.5.2 Number of test temperatures

The test temperatures at which calibration is to be carried out are dependent on the type and range of the thermometer. Generally about five test temperatures over the working range are adequate. The exact temperatures of calibration are specified for thermometers conforming to a particular specification. For example, the calibration temperatures for ASTM thermometers are given in ASTM E1 specification.

7.7.6 Calibration equipment

7.7.6.1 Ice point bath

An ice point bath is a convenient method of checking the zero of a thermometer. The ice point is the equilibrium temperature between ice and air saturated water.

An ice point bath can be easily constructed using a wide mouthed dewar flask, a siphon tube, ice and distilled water. A flask of diameter 80 mm and depth of about 400 mm would be adequate for most purposes, Fig. 7.10.

The ice is made from distilled or deionized water and finely crushed to small chips measuring 2 mm–5 mm. The flask is about one-third filled with

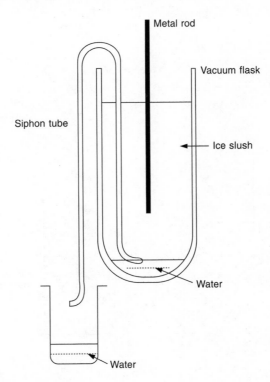

Figure 7.10 Ice point bath

distilled water and the crushed ice is added. The mixture is compressed to form a tightly packed slush, drained and remixed until there is no free water left, but the ice is completely wet. The ice will look glassy when it is completely wet. A siphon is placed in the flask to remove excess water formed as the ice melts.

The bath mixture is allowed about 30 minutes to attain temperature uniformity throughout before it is used. Ice should be added and excess water removed while the bath is being used. If precautions are taken to prevent contamination of ice and water, the ice point can be realized to better than ±0.01°C.

Commercial versions of ice point baths with an integral electrically operated stirrer are also available.

7.7.6.2 Stirred oil bath

Stirred oil baths are suitable for comparison calibrations in the range –30°C to +300°C. Beyond this temperature range baths using other media are used.

There are two main types of stirred oil baths, concentric tube type and parallel tube type. The schematic of a concentric tube type bath is shown in Fig. 7.11. The bath consists of two concentric cylindrical chambers. Fluid is filled in both chambers and is made to flow through the inner and outer

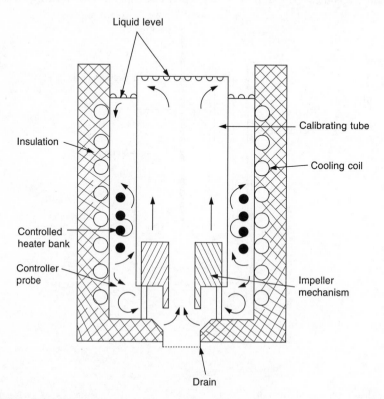

Figure 7.11 Schematic of concentric tube type bath

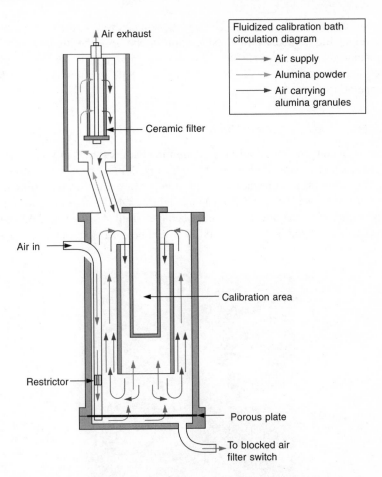

Figure 7.12 Schematic of a fluidized alumina bath. (Source: Isothermal Technology Ltd, UK)

chambers using a pump, which is fitted at the bottom of the inner chamber. A bank of heaters is used to maintain the temperature of the bath fluid at a set level or some baths incorporate a control system to raise the temperature slowly at a desired rate.

Two main criteria in choosing a bath are its *stability* and *uniformity*. Stability is the maximum variation of temperature of the fluid at the maximum operating temperature after stabilization of the bath. Uniformity is defined as the maximum difference in temperature at two different points in the bath at the same instant of time. The very best baths commercially available can achieve ±0.001°C for both stability and uniformity.

Different fluids are used depending on the temperature range required. Usually mineral or silicone oils are used in the range 20°C to 250°C. In the negative temperature range a number of other fluids including ethylene glycol (up to –30°C) are used. The boiling point as well as the flash point of the oil should be taken into consideration in choosing the higher operating temperature. Also the viscosity of the oil comes into play at the lower end of the range as some oils will get very thick and will not flow readily at low temperatures.

7.7.6.3 Fluidized alumina bath

The basic construction of a fluidized alumina bath is shown in Fig. 7.12 and a commercially available bath is shown in Fig. 7.13.

The bath consists of a container of aluminium oxide powder sitting on a porous base plate. Pressurized air is passed through the base plate to impart kinetic energy to the powder so that it behaves similarly to a fluid. When the powder is fluidized it displays good flow and heat transfer characteristics.

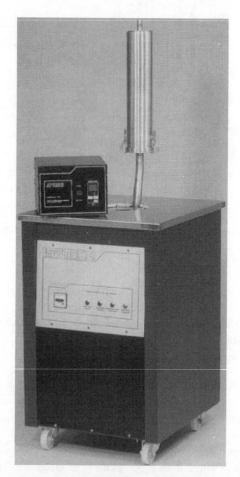

Figure 7.13 Fluidized alumina bath. (Source: Isothermal Technology Ltd, U.K)

However, the fluidized medium itself cannot achieve very good temperature stability and uniformity. Metal blocks are used to improve on these parameters. Fluidized baths operable in the range 50°C to 700°C with stability not exceeding ± 0.05°C are available commercially.

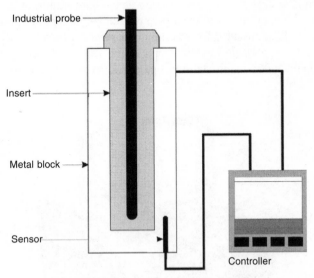

(a) Metal block calibrator of poor design

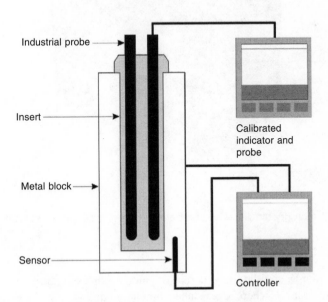

(b) Well-designed metal block calibrator

Figure 7.14 Schematics of dry block calibration baths. (Source: Isothermal Technology Ltd)

7.7.6.4 Portable calibrators

Dry block calibration systems are very convenient portable temperature calibrators. A number of different designs are available. Two common systems are illustrated in Fig 7.14.

Design A is considered a poor design as the temperature of the block is measured at a point far away from the probe under calibration. In design B, a separate hole is provided for insertion of a standard thermometer probe.

Baths having a range of –30°C to 1100°C are commercially available. Most manufacturers claim stabilities of the order of ±0.01°C up to 700°C (Fig. 7.15).

Figure 7.15 A dry block calibration bath. (Source: Isothermal Technology Ltd)

7.7.6.5 Simulators

Thermocouple simulators generate a voltage corresponding to an input temperature level, whereas resistance simulators provide a resistance output corresponding to an input temperature. Simulators can be used conveniently to verify the reading of display units of thermocouple and resistance thermometers, particularly when they are used in conjunction with a temperature transmitter. It is vital to remember that the simulator does not calibrate or verify the output of the sensor, calibration of which is vital to attain the required accuracy.

Generally thermocouple simulators are based on the electromotive force (emf) values given in reference tables for the letter designated thermocouple types. Resistance simulators are based on the resistance variation of the Pt-100 RTDs specified in the international standard IEC 60751.

Bibliography

International and national standards

1. ANSI MC96. 1. 1982. Temperature measurement thermocouples. American National Standards Institute.
2. IEC 60584–1. 1995. International thermocouple reference tables. International Electrotechnical Commission.
3. IEC 60584–2. 1982. Thermocouple tolerances. International Electrotechnical Commission.
4. IEC 60584–3. 1989. Extension and compensating cables. International Electrotechnical Commission.
5. IEC 60751. 1983. Industrial Platinum Resistance Thermometer Sensors. International Electrotechnical Commission.
6. ISO 4795 (1996). Glass for thermometer bulbs. International Organization for Standardization.
7. BS 1900 (1976). Specification for secondary reference thermometers. British Standards Institution.
8. ASTM-E1: 93–standard specification for ASTM thermometers. American Society for Testing and Materials.
9. BS 1704: 1985. 1992. Solid stem general-purpose thermometers. British Standards Institution.
10. BS 593: 1989. 1994. Laboratory thermometers. British Standards Institution.
11. ANSI/ASME B40.3. Bimetallic thermometers. American National Standards Institute/American Society of Mechanical Engineers.

Introductory reading

1. Beavis, M. (1983) *Techniques for the calibration of liquid in glass thermometers*, Commonwealth Scientific and Industrial Research Organization, Australia.
2. White, D.R. and Nicholas, J.V. (1994) *Traceable Temperatures*, Wiley Interscience.
3. Kerlin, T.W. and Shepherd, R.J. (1982). *Industrial Temperature Measurement*. Instrument Society of America.

Advanced reading

1. Preston, H. and Thomas, H. (1990). The International Temperature Scale of 1990, *Metrologia*, 27, 3–10, ibid. p. 107.
2. The International Practical Temperature Scale of 1968. Amended edition of 1975 (1976), *Metrologia* 12, 7–17.

8

Electrical measurement standards

8.1 Introduction

Measurement of electrical current, potential difference, resistance, capacitance and inductance are of vital importance in industrial processes. A number of different types of electrical measuring instruments such as multimeters, bridges, insulation testers, high voltage meters as well as resistors, capacitors and inductors are in use. A number of other equipment, e.g. electronic balances, load cells and analytical instruments, also use electrically operated transducers or sensors.

8.2 SI units

8.2.1 The ampere

The ampere is the SI base unit for electrical current. It is defined as:

> that constant current which, if maintained in two straight parallel conductors of infinite length, of negligible circular cross section, and placed one metre apart in vacuum, would produce between them a force equal to 2×10^{-7} newton per metre of length.

8.2.2 The volt

The volt is the SI unit of electrical potential difference and is defined as:

> the potential difference between two points of a conducting wire carrying a constant current of 1 ampere, when the power dissipated between these points is equal to 1 watt.

In terms of mechanical units the volt is:

$$1 \text{ volt} = \frac{1 \text{ watt}}{1 \text{ ampere}}$$

$$1 \text{ watt} = \frac{1 \text{ joule}}{1 \text{ second}}$$

$$1 \text{ joule} = 1 \text{ newton} \cdot 1 \text{ metre}$$

8.2.3 The ohm

The ohm, the unit of electric resistance, is defined as:

the electric resistance between two points of a conductor when a constant potential difference of 1 volt, applied to these points, produces in the conductor a current of 1 ampere, the conductor not being the seat of any electromotive force.

8.2.4 The farad

The farad is the SI unit of capacitance. It is derived from the volt, ampere and second as:

$$1 \text{ farad} = \frac{1 \text{ coulomb}}{1 \text{ volt}}$$

$$1 \text{ coulomb} = 1 \text{ ampere} \times 1 \text{ second}$$

The coulomb is the SI unit of electric charge.

8.2.5 The henry

The henry is the SI unit of inductance. It is derived from the volt, second and the ampere as:

$$1 \text{ henry} = \frac{1 \text{ weber}}{1 \text{ ampere}}$$

$$1 \text{ weber} = 1 \text{ ampere} \times 1 \text{ second}$$

The weber is the SI unit for magnetic flux.

The above SI definitions were adopted by the 9th CGPM held in 1948, and are still valid. The realization of the units using physical standards gives rise to numerous practical difficulties. Thus the ampere, the volt, the ohm and other units realized by the BIPM and various national laboratories are quoted with uncertainties arising from inherent limitations of the physical standards and measurement conditions used.

8.3 Primary standards

8.3.1 Current

The current balance is the primary standard for the realization of the ampere. A current balance uses the principle enunciated in the definition of the ampere, though in a more practical way. Instead of using long, straight conductors, two circular coils are used. One coil is fixed and the other is mounted at the end of a balance arm. The force generated between these coils when a constant current is passed through them is counterbalanced by known weights placed on the other arm of the balance. A schematic of the current balance is shown in Fig. 8.1.

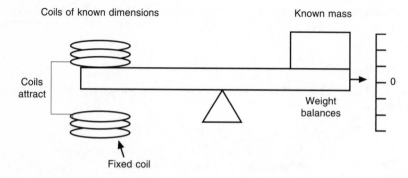

Figure 8.1 Current balance

The uncertainty of realization of the ampere using a current balance is about 15 parts per million (ppm). Due to the difficulty of realization of the ampere using a current balance and the availability of primary standards for the volt and the ohm, in many local and national laboratories it is derived from the ratio of volt and the ohm, using Ohm's law:

$$1 \text{ ampere} = \frac{1 \text{ volt}}{1 \text{ ohm}}$$

8.3.2 DC voltage

8.3.2.1 Josephson standard

Brian Josephson, working at the then National Bureau of Standards (NBS), discovered the Josephson effect in 1962. A Josephson junction consists of two superconductors, separated by a thin oxide insulating barrier. When a DC current biased junction is irradiated with a microwave source a DC voltage proportional to the frequency of the microwave source is produced across the junction.

A series of constant DC voltages appears across the junction due to quantum mechanical tunnelling of an AC current through the junction. The DC voltage steps are given by the equation:

$$V = \frac{fnh}{2e} = \frac{fn}{K_j}$$

where:
f = frequency of the AC current
n = step number
h = Planck's constant
e = electronic charge value.

K_j is known as the Josephson constant and has an assigned value of 483 597.9 GHz/V.

The junction has to be maintained at a temperature below 4.2 kelvin and is located in a dewar containing liquid helium. The voltage produced by a single junction is of the order of 10 mV. Present day Josephson standards consist of arrays of junctions connected in series producing as much as 10 V DC output, Fig. 8.2. Recent experiments have shown that the DC voltage developed by Josephson junctions irradiated by a common source of microwaves agrees to within 3 parts in 10^{19}.

Researchers working at the National Institute of Standards and Technology (NIST) have described a compact, transportable and fully automated Josephson

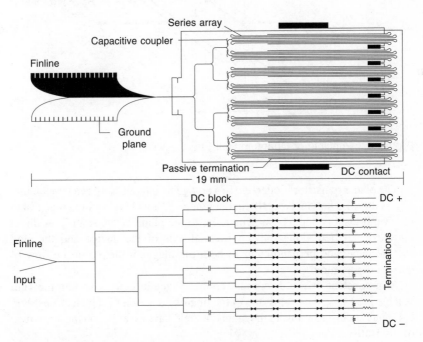

Figure 8.2 Josephson array. (Source: Fluke Corp.)

standard recently. The uncertainty of realizing the volt using this standard is reported to be 1 part in 10^9 at 95 per cent confidence level.

8.3.3 Resistance

8.3.3.1 Quantized Hall resistance standard

The quantized Hall resistance standard is based on the Quantum Hall Effect (QHE). Klaus von Klitzing observed the Quantum Hall Effect in 1980 in high quality silicon MOSFETs (Metal Oxide Semiconductor Field Effect Transistors). Subsequent experiments in collaboration with the Physikalisch Technische Bundesanstalt (PTB) in Germany and workers in several other national laboratories confirmed that the Quantum Hall Effect could be used to construct an international resistance standard. A planar MOSFET transistor constructed to constrain a current of electrons within a thin layer is the main component of a QHE device. The schematic of a simplified QHE device is shown in Fig. 8.3.

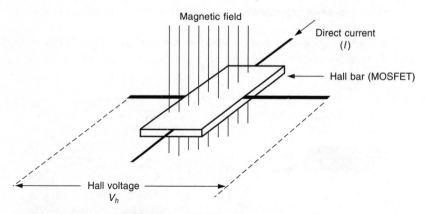

Figure 8.3 Principle of Quantum Hall Effect device

The planar transistor is operated in a cryogenic environment at a temperature of less than 4.2 kelvin, typically in the range of 1 to 2 kelvin. A magnetic field of several tesla is applied perpendicular to the plane of the sample. A direct current is applied along the longitudinal axis of the device and the Hall voltage generated perpendicular to both the magnetic field and the flow of current is measured.

At a constant drive current and as the magnetic field is varied, the Hall voltage is observed as steps due to the quantized action of the electrons, Fig. 8.4. Thus small variations in the magnetic field have no effect on the amplitude of the Hall voltage.

The Hall voltage and the Hall resistance are given by:

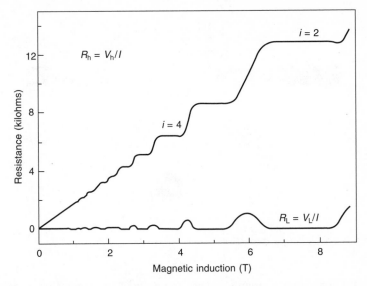

Figure 8.4 Variation of the Hall (R_h) and the longitudinal (R_L) resistance with magnetic field. (Source: National Research Council, Canada)

$$V_h = R_{k-90}\, I/n$$

$$R_h = V_h/I = R_{k-90}/n$$

where:

V_h = Hall voltage

R_{k-90} = assigned value of von Klitzing constant

R_h = Hall resistance

n = an integer representing the plateau where the value of R_h is measured.

The recommended international value of the von Klitzing constant is 25 812.807 ohms exactly.

Portable Hall resistance standards are also becoming available. A modular and portable standard has been recently reported by researchers of the Institute for National Measurement Standards of the National Research Council of Canada.

8.3.4 Capacitance

8.3.4.1 Calculable capacitor

The SI unit of capacitance, the farad, is realized by using a Thompson-Lampard calculable capacitor. The calculable capacitor is a linear device known as a cylindrical cross–cross capacitor. The capacitance per unit length of a cross capacitor can be computed with great precision. A simplified diagram of a cross capacitor is shown in Fig. 8.5.

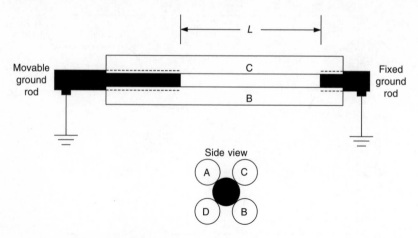

Figure 8.5 Calculable capacitor

In this simplified diagram four bars of circular cross-section are placed so that they are at the vertices of a rectangle when viewed end on. The four parallel bars are enclosed in an electrostatic shield. The two diagonally opposite sets of bars each constitute one capacitor.

If the capacitance of the two sets of opposite bars are designated C_1 and C_2, then according to Lampard the mean capacitance $\overline{C}$ is given by:

$$\overline{C} = \left(\frac{C_1 + C_2}{2} \right)$$

and

$$\overline{C} = \varepsilon_0 L \left[\frac{\ln 2}{\pi} \right] \left[1 + 0.087 \left(\frac{(C_2 - C_1)}{\overline{C}} \right)^2 \right]$$

$$+ \text{ fourth order and higher order terms}$$

where:
$\varepsilon_0 = 1/\mu_0 c^2$
L = length of bars in metres
μ_0 = permeability of vacuum
c = speed of light in vacuum

if $C_1 = C_2 = C$ then $C = \varepsilon_o L (\ln 2)/\pi$

Usually the fourth and higher order terms can be neglected and the capacitance is approximately 2 pF/m.

However, errors arise due to distortion of the electrical field near the ends of the bars. A number of techniques such as movable ground rods are utilized to overcome these difficulties.

8.4 Secondary standards

8.4.1 Standard cells

A bank of saturated Weston cadmium cells (Fig. 8.6) had been the DC voltage standard maintained by standards laboratories for a number of years. The saturated cell when carefully used exhibits good voltage stability and long useful life.

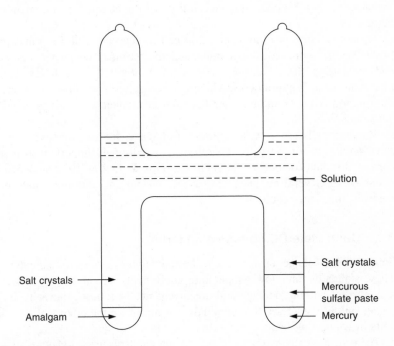

Figure 8.6 Weston cell

The nominal output voltage of a saturated cell is 1.0183 volts at 20°C. Individual cells may differ from the nominal by tens of microvolts. The saturated cell has a relatively large temperature coefficient approximately −40 ppm/°C at 20°C and therefore has to be kept in a constant temperature bath or oven to maintain a stable voltage.

Standard cell enclosures available commercially usually contain four cells and operate at about 30°C. Substantial hysteresis is displayed by a cell if its temperature is changed and then returned to normal. Following a change in temperature a cell will either return to its previous value or to a new value. However, the stabilization period to reach within 0.5 ppm of its previous value can be as much as 90 days. Shock or vibration, tripping, gas bubble formation at the electrodes or current flow into or out of the cell may also cause a cell to drift appreciably for extended periods of time.

8.4.1.1 Maintenance of standard cells

Standard cells can exhibit unpredictable drifts or shifts in their output voltage. For this reason a number of four cell enclosures constituting a bank are used to maintain a voltage standard. At least three four cell enclosures (12 cells) are needed for a reliable voltage standard. The voltage outputs of individual cells are intercompared regularly to determine their deviations from the mean value of the bank with the measurement results plotted on control charts. This arrangement also allows one (or two) enclosures of the bank to be used as a transfer standard to maintain traceability with a national or international laboratory.

The comparative measurements of the cell voltages are made by connecting two cells together in series-opposition, negative terminals connected together, and measuring the difference with a potentiometer or high resistance microvoltmeter. Test protocols, which make it possible to obtain the desired information with a minimum number of measurements, are given in NBS Technical Note 430.

Standard cells are very susceptible to damage due to excessive current flow into or out of the cells. Therefore it is extremely important to avoid current flow into or out of the cells. Currents as low as 10^{-15} amperes, if sustained for several minutes, can cause a change in value for which long recovery times are required.

8.4.2 Solid state DC voltage standards

Solid state DC voltage standards are widely used today in standards laboratories. These standards have low temperature coefficients, are able to source and sink current without damage and are mechanically robust. They also have higher output voltages (10 V nominal) that minimize the effects of thermal emfs in connecting leads.

There are basically two types of DC voltage standards in use today: *reference amplifier type* and *discrete Zener diode type*.

8.4.2.1 Reference amplifier type

A reference amplifier is an integrated circuit consisting of a Zener diode and a transistor and is illustrated in Fig. 8.7.

There are two advantages of using a reference amplifier. The current in the Zener diode can be set independent of the transistor's base current. This allows the amplifier's collector current to be adjusted so that the temperature coefficient of the output is near zero, over a narrow temperature range. Also the reference amplifier's integral transistor allows it to be used in association with a high gain, negative feedback amplifier to provide a regulated output at a higher reference voltage.

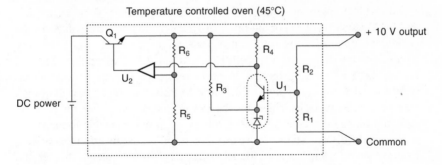

Temperature controlled oven (45°C)

Figure 8.7 Reference amplifier. (Source: Fluke Corp., USA)

8.4.2.2 Discrete Zener diode type

A reverse biased diode known as the Zener diode has been in use as a voltage regulator from the advent of solid state devices in the 1950s.

Early Zener diodes were not very stable and had relatively high noise levels. Today's improved Zener diodes are as stable as saturated standard cells. They are also mechanically robust, unaffected by reasonable levels of shock and vibration and relatively stable to extreme temperature variations.

Figure 8.8 shows a Zener diode combined with a forward biased diode to achieve near zero temperature coefficient. In this arrangement the temperature coefficient of the Zener (2 mV/°C), which is positive, is compensated by the nearly equal negative temperature coefficient of the forward biased diode (−2 mV/°C) to yield a device having a near zero temperature coefficient.

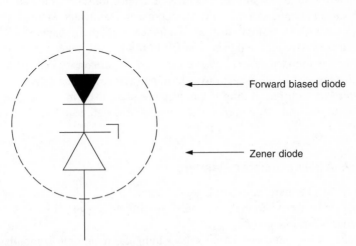

Forward biased diode

Zener diode

Temperature coefficient (TC) of the forward biased diode is approximately equal to but of opposite polarity of Zener TC

Figure 8.8 Discrete Zener standard (Source: Fluke Corp., USA)

The typical operating voltage of a Zener reference diode is between 6.2 V and 6.3 V. However, 10 V, or multiples of 10 V, are the preferred values for the standards laboratory. Most solid-state voltage references used in the laboratory are designed to deliver 10 V, or both 10 V and 1.018 V. These higher voltages are obtained by feeding the 6.2 V Zener reference voltage to a regulated precision 10 V power supply. Isolation for the Zener reference as well as high current capability is provided by the power supply amplifier. In addition current limiting for accidental short circuits is also provided in this arrangement.

The main disadvantage of the discrete amplifier is uncertainties arising from offset drift and noise in the amplifier, drift in the voltage divider and drift in the output adjustment control adding to the drift in the Zener reference voltage. If the standard has 1 V or 1.018 V outputs obtained by dividing the 10 V output then these will be subject to additional uncertainties arising from the drift in the resistive voltage dividers.

8.4.2.3 Maintenance of solid-state standards

A solid-state voltage standard should always be maintained under power as it may exhibit a small change in output when power is removed and reapplied. This is especially important when solid-state references are shipped.

8.4.3 Stability of output with time

The stability of the 10 V output of a solid-state standard is about 0.5 ppm/ year. Unlike in the case of standard cells where at least 12 cells are needed to maintain a reliable standard, about four solid-state standards are sufficient to maintain a voltage standard. The cells of a bank are intercompared regularly using a high impedance digital voltmeter or a potentiometer. Many laboratories have automated systems using a computer, a thermal scanner and digital voltmeter interconnected using the IEEE 488 interface bus.

Such automated systems enable the intercomparisons of cells on a regular basis (say every two weeks) to determine their drift characteristics. Linear regression and trend fitting techniques are often used to predict the output at a future time. A commercially available 4-cell standard is shown in Fig. 8.9.

8.4.4 AC–DC transfer standard

An AC–DC transfer standard is used to transfer DC voltages and currents to AC quantities or vice versa. A number of different models of AC–DC transfer standards are in use today.

The older standards based on vacuum thermocouple principles were developed from work done at NBS (presently NIST) in the 1950s. Most of the current types of AC–DC standards use a solid-state device known as RMS sensor introduced in 1974.

Figure 8.9 Fluke 734A solid-state standard. (Source: Fluke Corp., USA)

8.4.4.1 Vacuum thermocouple-based standard

A vacuum thermocouple-based standard uses a thermal element that works on the principle of heat generated in a resistance sensed by a thermocouple attached to it. Since the heat generated in a resistive element is only dependent on the mean current passed through it, the root mean square (RMS) value of an alternating current can be made equal to the value of a direct current passed through the element.

The main disadvantage of a single thermocouple is that its output is only a few millivolts (typically 5–10 millivolts). The other problems are their long thermal time constant and direct voltage reversal error. In later years the introduction of multijunction thermal elements giving outputs of the order of 100 millivolts overcame the problem of low voltage output. However, the long thermal time constant, direct voltage reversal error and poor frequency response (above 100 kHz) were not entirely overcome.

8.4.4.2 RMS sensor-based standard

The original RMS sensor introduced in 1974 was a monolithic device having a transistor and a diffused resistor on a silicon substrate. Power dissipated in the resistor is detected by the transistor through the temperature sensitivity of the base-emitter junction voltage, Fig. 8.10.

This device overcame the major disadvantage of the thermocouple or multijunction thermal element, namely the long thermal time constant. However, this sensor had a direct voltage reversal error as large as, if not larger than, the single junction thermocouple.

These problems were overcome by constructing the RMS sensor as a differential pair using two resistor transistor combinations, Fig. 8.11. The application of an input voltage to R_1 causes the temperature of the base emitter junction of T_1 to change. This causes a change in base-emitter voltage (V_{be}) of T_1 which in turn controls the collector voltage. The differential amplifier A compares the collector voltage of T_1 and T_2. By driving a current through R_2 the base-emitter voltage of T_2 is changed causing a change in the collector

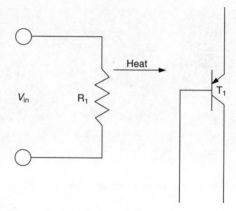

Figure 8.10 Thermal sensor. (Source: Fluke Corp., USA)

voltage. A balance occurs when the collector voltages of T_1 and T_2 are equal. When this happens the input voltage V_{in} of any waveform is approximately equal to the direct voltage at V_{out}.

8.4.5 Resistance

8.4.5.1 Thomas one ohm standard

The Thomas type one ohm standard is a highly stable wire wound resistance standard. The resistance element is contained in a hermetically sealed, double

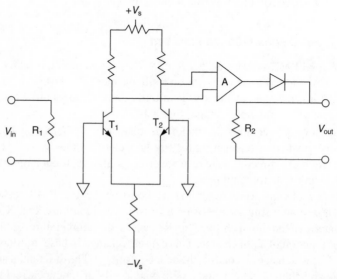

Figure 8.11 Differential pair RMS sensor. (Source: Fluke Corp., USA)

walled metal container. The temperature coefficient of resistance of these resistors is relatively high and therefore they have to be maintained in a temperature controlled oil bath.

The Thomas type one ohm resistance standard is widely used as a secondary reference standard for resistance in many standards laboratories. The value of one ohm standards is transferred to higher valued standards using bridge comparison methods.

The drift of a well-maintained Thomas type resistor is of the order of 0.02 ppm per year.

8.4.5.2 CSIRO one ohm standard

More recently, one ohm type standard resistors having superior characteristics to those of Thomas type, especially with respect to temperature and pressure coefficients, have been developed by the Commonwealth Science and Industrial Research Organization (CSIRO) of Australia.

The resistive element is a coil of Evanohm wire annealed at 980°C. Further heat treatments at lower temperatures are used to obtain a very small temperature coefficient (1×10^{-6}/K or lower at 20°C) at ambient temperatures. The heat-treated coil is mounted in a special support and is unsealed. The pressure coefficient is negligible and long-term stability is a few parts in 10^8 per year.

8.4.5.3 Reichsanstalt artefact standard

Reichsanstalt design is often used for low valued standard resistors, from 0.001 ohms, to 0.1 ohms which are subjected to high current in use. In this design the resistance wire is wound on a metal tube, which is enclosed in a perforated container. When immersed in a temperature controlled oil bath, the perforations allow free circulation of oil over the resistance element, thereby aiding in cooling of the resistor, particularly when high test currents are used.

8.4.5.4 Rosa artefact standard

A design proposed by Rosa is frequently used for construction of standard resistors of value 10 ohms and higher. This design is also known as NBS type (Fig. 8.12).

A resistance element wound on an insulated metal tube is suspended in an oil filled outer can. A concentric inner tube supports the insulated resistor form and provides a thermometer well for insertion of a thermometer to determine the resistor ambient temperature.

The oil in the can provides heat transfer from the resistance element to the outer wall of the can. The whole container is also immersed in a temperature controlled oil bath for further stabilization of the temperature.

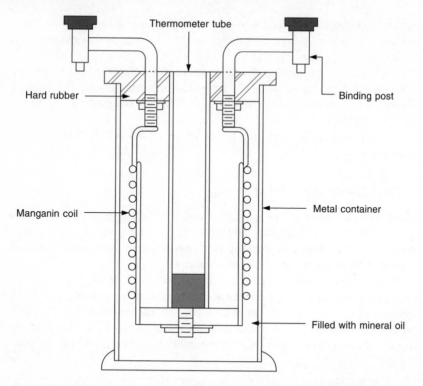

Thermometer tube

Hard rubber

Binding post

Manganin coil

Metal container

Filled with mineral oil

Figure 8.12 NBS one ohm type standard. (Source: National Institute of Standards and Technology, USA)

8.4.5.5 ESI SR IO4

ESI SR 104 is a 10 kilohm transportable resistance standard manufactured by Electro Scientific Industries. The ESI SR 104 is widely used in standards laboratories. It is also used in interlaboratory comparisons of resistance due to its portability. The standard has its own oil bath.

The long-term stability of the standard is specified as ±1 ppm in the first year and ±0.5 ppm/year after 2 years. The temperature coefficient is ±0.1 ppm/°C at 23°C. A built-in temperature sensor that can be used to measure the internal temperature of the resistor is provided. A table of correction factors to correct the resistance value to the local ambient temperature is also provided.

8.4.6 Capacitance

A number of different types of secondary standard capacitors are available. Those in the range 1 pF to 1000 pF are constructed with a solid dielectric or with an inert gas dielectric. Capacitors of 10 pF or 100 pF with solid dielectric (fused silica) are available. Gas dielectric capacitors of 5 pF and 10 pF are also used. Gas dielectric capacitors of higher value, 100 pF to 1000 PF are made of piles of separated parallel plates in a hermetically sealed container with nitrogen as the dielectric.

Standard capacitors usually have either three terminals or five terminals. In the three terminal type a Faraday shield is used. The shield is connected to the third terminal. A schematic of the three terminal type capacitor is shown in Fig. 8.13.

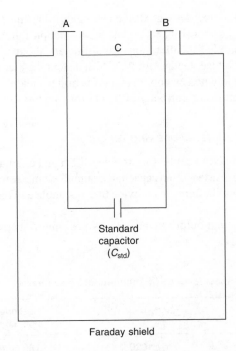

Figure 8.13 Three terminal type capacitor. (Source: Fluke Corp., USA)

The two additional terminals of a five terminal capacitor enable a four terminal connection to be made, that is, one pair of terminals is used for current application and the other pair for measurement of voltage. The shield is used as in the case of a three terminal capacitor.

In parallel plate capacitors, a guard ring is used to reduce fringing, which occurs at the outer edges of the active plates. Generally secondary standard capacitors should be calibrated against a primary capacitance standard every two years or so.

8.4.7 Inductance

Although it is possible to construct calculable inductance standards by winding a conductor on an insulated core, this is generally not done as it is more prudent to construct accurate standard capacitors and assign values of inductance to them.

8.5 Working standards

8.5.1 Multifunction calibrator

The workhorse of the present day electrical calibration laboratory is the multifunction calibrator (Fig. 8.14). The multifunction calibrator is a working standard used mainly for calibration of multimeters, though some models can be used as a DC voltage standard for AC–DC transfer measurement of alternating voltages or as a low noise supply source for bridges and dividers.

The main features of currently available instruments are given here.

8.5.1.1 Calibrator functions and ranges

A typical multifunction calibrator provides direct and alternating voltage and current as well as resistance and capacitance stimuli. Some instruments marketed as multiproduct calibrators also provide thermocouple and resistance thermometer stimuli.

The functions and output ranges of a typical multifunction calibrator are given in Table 8.1.

Table 8.1 Typical specifications of a multifunction calibrator

Function	Range	Best one year specification ppm or % of setting
DC voltage	0 to ±1020 V	±50 ppm
DC current	0 to ±11 A	±0.01%
Resistance	0 to 330 MΩ	±0.009%
AC voltage 10 Hz to 500 kHz, sine	1 mV to 1020 V	±0.03%
AC current 10 Hz to 10 kHz, sine	29 μA to 11A	±0.06%
Capacitance	0.33 nF to 1.1 mF	±0.25%
Thermocouple	−250°C to +2316°C	±0.14°C
RTD source	−200 to +630°C	±0.03°C
DC power	109 μW to 11 kW	±0.08%
AC power 45–65 Hz, PF = 1	109 μW to 11 kW	±0.15%
Phase 10 Hz to 10 kHz	0 to ±179.99°	±0.15°
Frequency	0.01 Hz to 2.0 MHz	±25 ppm

8.5.1.2 Output drift

Most components of electrical instruments are subject to change due to effects arising from loading and ambient conditions (temperature, humidity, etc.). This phenomenon is known as electrical drift. For this reason, the specifications of a calibrator are usually given for a time period. Ninety days or one year specifications are common. This means that the stimuli of the calibrator are expected to be within the tolerance indicated during the time period given. At the end of the time period a comparison against a higher level standard is required to determine the relative drift of the stimulus. Good quality calibrators should have small drift rates over a considerable period of time. Typical specifications of a modern multifunction calibrator are given Table 8.1.

8.5.1.3 Burden current

A calibrator operated in voltage mode has to supply a burden or load current. Since the input impedance of a modern digital multimeter (DMM) is typically 10 MΩ or higher in the DC voltage ranges the burden current would be limited to about 0.1 mA at 1000 V. In the AC voltage range and for certain other applications burden currents of 20 mA or more may be required.

8.5.1.4 Compliance voltage

The voltage required to drive a constant current through the unit under test is known as the compliance voltage. In the case of DMM calibrations this voltage is the sum of the voltages across the DMM shunt (see Fig. 8.16), lead resistances and contact resistances of the binding posts.

8.5.1.5 Protection

Protective circuits are incorporated in all well-designed calibrators for guarding against erroneous connections. The calibrator should be able to tolerate a potentially dangerous short circuit or any other incorrect connection by switching to standby mode.

8.5.1.6 Output noise

The output noise is a factor that must be taken into consideration in the selection of a calibrator. On a given range the output noise of the calibrator must be low enough to keep the least significant digit of the unit under test from being affected.

8.5.2 Process calibrator

A process calibrator is a portable instrument intended mainly for use in the field for troubleshooting and calibration of process control instrumentation.

Figure 8.14 Fluke 5520A multifunction calibrator. (Source: Fluke Corp., USA)

These rugged, battery operated calibrators are available as single function or multifunction instruments. The typical functions and specifications of a process calibrator are given in Table 8.2.

Table 8.2 Functions and specifications of a typical process calibrator

Function	Resolution	Accuracy (1 year) Source	Measure
DC voltage			
0 to 100 mV	0.01 mV	0.02% Rdg + 2 LSD	0.02% Rdg + 2 LSD
0 to 10 V	0.001 V	0.02% Rdg + 2 LSD	
0 to 30 V	0.001 V		0.02% Rdg + 2 LSD
DC current			
0 to 24 mA	0.001 mA	0.02% Rdg + 2 LSD	0.02% Rdg + 2 LSD
Resistance			
0 Ω to 3200 Ω	0.01 Ω to 0.1 Ω		0.10 Ω to 1.0 Ω
15 Ω to 3200 Ω	0.01 Ω to 0.1 Ω	0.10 Ω to 1.0 Ω	
Frequency			
1 to 1000 Hz	1 Hz	0.05% Rdg	
1.0 to 10.0 kHz	0.1 kHz	0.25% Rdg	
Thermocouples			
Types J, K, T, E, N	0.1°C	0.7°C–1.2°C	0.7°C–1.2°C
Types R, S, B,	1°C	1.4°C–2.5°C	1.4°C–2.5°C
Resistance temperature detectors (RTD)			
Pt 100		0.3°C	0.3°C
Ni 120		0.2°C	0.2°C

LSD–least significant digit.
Rdg–reading.

8.5.3 Resistors

Present day working standard resistors usually do not require oil or air baths and therefore can be used for on-site calibrations as well. Wire wound working standard resistors in the range of 10 Ω to 20 MΩ are available. Most resistors have temperature stability of typically less than 2 ppm over a temperature interval of ten degrees (18°C to 28°C). In most cases manufacturers supply calibration tables, listing corrections in 0.5°C increments. In good quality resistance standards permanent shift in resistance is typically less than 2 ppm after cycling between 0°C and 40°C.

8.6 Calibration of a multifunction calibrator

There are two aspects to calibration, *verification* and *adjustment*. These are used to determine and correct for the classes of errors that a calibrator can exhibit. In some environments the item to be calibrated is verified when it is returned to the calibration laboratory at the end of the recalibration cycle and when it leaves the laboratory after it has been adjusted. These two verifications are sometimes referred to as *as-found* and *as-left* calibration.

8.6.1 Verification of calibrator

During verification, the calibrator's outputs on each range and function are determined by comparison to known standards but are not adjusted. A comprehensive set of measurement standards is usually required to perform these tests.

8.6.2 Adjustment of calibrator

The adjustment of the calibrator is undertaken when the outputs are found to be outside of the specification limits. There are three available methods of adjustment:

(a) Artefact calibration adjustment.
(b) Software calibration adjustment.
(c) Manual calibration adjustment.

The most advanced of these techniques is artefact calibration. It uses a small number of single-valued artefacts to provide a fully traceable calibration using a complement of calibration standards. The calibrator's microprocessor senses the difference between its output and the standard value and stores it in firmware. This is applied as a correction to the calibrator's output.

Software calibration is done in much the same way as a manual calibration. A correction factor for its outputs on each range and function is obtained by comparison to external standards. During operation, the microprocessor applies this correction factor to the calibrator outputs, e.g. Fluke 5440B and 5450A models are adjusted using this technique.

In a manual adjustment, the calibrator outputs are adjusted against external standards through internal controls. The Fluke 5100B and other earlier models are adjusted in this manner.

8.7 Calibration of multimeters and other instruments

8.7.1 Reference standard

The hierarchy of calibration standards used for calibration of electrical measuring instruments and components is given in Fig. 8.15.

Figure 8.15 can be used as a guide for deciding on the next higher level standard against which a given unit should be calibrated. Primary standards

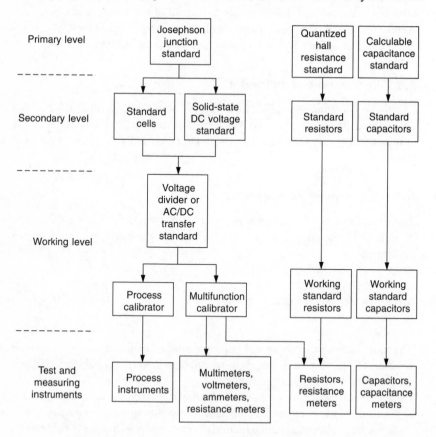

Figure 8.15 Hierarchy of electrical measurement standards

are those described earlier in the chapter namely the Josephson voltage standard, the Hall resistance standard and the calculable capacitance standard. Secondary standards are those calibrated against the primary standards, namely standard cells and solid-state DC voltage standards, secondary standard resistors and secondary standard capacitors.

Working standards are the AC–DC transfer standard, voltage divider, multifunction calibrator, process calibrator, working standards of resistance, and working standards of capacitance and a number of other equipment used as secondary standards for different parameters.

If the technical manual of the instrument is available with the adjustment potentiometers and trimmers clearly identified then it is possible to adjust the instrument to give minimum deviations. The adjustment is usually done at the middle of the scale or at a designated voltage or current level indicated in the manual. If the technical manual is not available then the deviations are recorded in a test report and corrections should be applied when using the instrument. However, this is hardly possible for instruments used in the field. As a general rule, instruments that are found to be out of specification and cannot be adjusted should be clearly marked as 'Out of Specs'.

8.7.2 Reference conditions

The reference temperature for electrical calibrations is 23°C. A laboratory with temperature controlled to about ±1°C is generally used. A low relative humidity of around 40–50 per cent is necessary since high humidity levels affect electrical components. One of the most important requirements would be to have an environment with minimal electrical and electromagnetic fields.

8.7.3 Calibration of multimeters

8.7.3.1 Analogue multimeters

Analogue multimeters are rather rare now. However, a few basic principles used in their calibration are included.

Zero adjustment

The zero of an analogue multimeter is adjusted with the use of a screwdriver. Care must be taken not to exert any undue force on the movement, as this would damage the fine spring attached to the moving coil.

Voltage and current ranges

The voltage and current stimuli are obtained from a voltage or current calibrator or multifunction calibrator. Three positions of each voltage and current range need to be verified. Usually readings are taken in the lower end, middle and higher end of the scale. At least three readings at each position should be taken and the mean value worked out.

Resistance ranges

Resistance ranges can be verified using the resistance references of a multi-function calibrator or by connecting working standard resistors at the terminals of the multimeter. However, adjustment of the resistance ranges of an analogue multimeter is quite difficult as the scales are not linear especially at the higher end of the scale.

8.7.3.2 Digital multimeters

There are a large variety of digital multimeters, the main types being *handheld*, *bench* and *laboratory* types. All these types are calibrated using a multifunction calibrator, although in the case of the laboratory type multimeters special precautions have to be taken as the test uncertainty ratio approaches 1:1.

Handheld type

The handheld digital multimeter is the most commonly used type and typically has a $3^1/_2$ or $4^1/_2$ digit display with five electrical functions, AC–DC voltage, AC–DC current and resistance. Most models also have frequency, continuity, diode test, and capacitance and waveform functions. A few enhanced models also have large LCD panels with graphics display ability. These instruments require regular calibration generally once in six months or a year depending on usage and the particular model.

Bench type

Bench type digital multimeters are used in many electronic workshops and test installations. These usually have $4^1/_2$ or $5^1/_2$ digit displays and smaller uncertainties than the handheld type. These also have five electrical functions and additional functions of frequency, capacitance, thermocouple inputs, etc. depending on the model. Some models also have RS-232 or IEEE-488 interfaces enabling them to be calibrated in closed loop configuration.

Laboratory type

Laboratory type is the most accurate type of digital multimeter available and their uncertainties approach those of the multifunction calibrator. These generally offer five functions AC–DC voltage, AC–DC current and resistance and are available with up to $8^1/_2$ digit display.

Laboratory type digital multimeters contain microprocessors and memory chips that allow them to perform complex mathematical computations and store corrections on all functions and ranges. These instruments are often calibrated in automated closed loop mode using their IEEE-488 interfaces.

8.7.3.3 General calibration techniques

Detailed calibration procedures are beyond the scope of this book. It is very important to follow the procedure indicated in the technical manual in performing the calibration of a particular instrument. However, some important procedures given in *Calibration: Philosophy in Practice* (Reference 1) is reproduced in this section with kind permission of the Fluke Corporation.

Digital multimeters of all types basically consist of two main sections, the measurement section and the control section as shown schematically in Fig. 8.16.

The measurement section is made up of the analogue conditioning circuits and analogue-to-digital converter. The control section consists of a microprocessor, memory and associated circuitry. The measurement section requires verification and calibration adjustment. The control section requires no calibration adjustment as it is usually checked internally on power-up.

The sections that require calibration adjustment are as follows.

Internal references

These are internal DC voltage and resistance references. In many instances a direct measurement of the DC voltage reference or resistance reference is not required. Instead a known input voltage or resistance is applied at the input and adjustment is made until the instrument displays the applied input voltage or resistance.

Analogue-to-digital converter (ADC)

Since all functions use the ADC, it must be calibrated as one of the first steps. Some digital multimeter calibration schemes call out separate ADC and reference calibrations; other schemes merely call out DC voltage range calibrations only. Digital multimeters whose procedures do not call out separate ADC adjustments do have internal corrections. The calibration procedure simply calls for an adjustment in the straight-in DC voltage range first. This range, going directly into the ADC, has unity gain and thus has no significant error in scale factor. The calibration procedure will often call for a zero offset adjustment of the DC range, and then the ADC is adjusted

DC voltage range

Digital multimeters usually have full-scale ranges of 200 V DC, 300 V DC, or 1000 V DC, typically divided by factors of 10 to determine the lower ranges. Normal calibration practice limits the testing of ranges to just less than full scale. This is because most digital multimeters are auto ranging, and

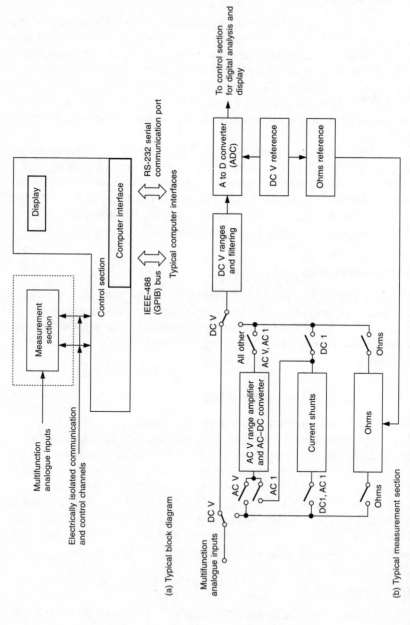

(a) Typical block diagram

(b) Typical measurement section

Figure 8.16 Sections of a digital multimeter. (Source: Fluke Corp., USA)

will go to the next higher range if the input is a certain percentage of full scale. So the calibration voltages for a digital multimeter with a full-scale range of 200 V DC usually follow a 190 mV, 1.9 V, 19 V and 190 V sequence. Modern multifunction calibrators take advantage of this structure.

The zero on each range may also need to be adjusted. The calibration procedure is to apply a short circuit to the input to the amplifier and adjust the digital multimeter reading to exactly zero volts DC. This would be followed by an input near full scale, in this case, 190 mV, DC. The digital multimeter is then adjusted to display the exact input voltage. In effect, the user is making the gain exactly 100. A digital multimeter calibration may also require an input of opposite polarity, –190 mV. This is to correct either for secondary linearity errors of the range amplifier, or for linearity errors within the ADC.

AC–DC converter

AC–DC converter calibration is similar for all types of converters. Amplitude adjustments are often made at two or more frequencies on each range. Zero adjustments are not usually made.

Average responding AC–DC converters

Calibration adjustment of average responding converters corrects for deviations in the equation in the transfer function equation, $y = mx + b$ (y = output, x = input, M = gain, b = offset). However, the b term is solved not for inputs at zero volts, but typically for inputs at one-tenth or one-hundredth of full scale. This is because precision rectifiers tend not to work well at zero volts input. Near zero volts, they rectify noise that would otherwise ride on the input waveform and average to zero. By moving the b term calibration off zero, this problem is avoided.

True RMS AC–DC converters

True RMS converters need similar calibration of ranges at full-scale and at near-zero input levels. But true RMS circuits have an even greater problem near zero volts than average responding converters. For example, analogue log/antilog schemes employ a precision rectifier, resulting in noise rectification problems like those associated with an average responding converter. Also, since the displayed output voltage is the root mean square voltage of the signal and noise voltages, very small input signals cause the noise voltage to dominate. Consequently, true RMS calibrations are accomplished with inputs well above zero volts.

There are additional adjustments to correct for errors in linearity of RMS converters. If the digital multimeter has a DC-coupled AC voltage function, there may be zero input adjustments to correct for offset errors of the scaling amplifiers.

The AC voltage converter can have a separate calibration for the true RMS converter, but often its calibration is combined into the main AC voltage function calibration. For example, if the true RMS module can handle a 2 V full-scale waveform, the 2 V AC range should be calibrated first; then adjust the true RMS module. The other ranges are then adjusted to correct for their gain errors. In some cases, the true RMS module needs more than two different inputs to correct for linearity errors. For example, an AC voltage calibration adjustment may call for inputs of 19 mV, 190 mV and 1.9 V on the 2 V range.

Scaling sections, averaging converters and true RMS converters have resistive and capacitive errors. A major consideration when calibrating AC voltage is frequency response. The general requirement is to use a constant-amplitude calibrator that covers the entire frequency range of the digital multimeter. The conversion takes some time, and slewing through the passband is time consuming. Because of this, spot frequency checks are usually made instead.

Low frequency calibration is generally performed at 400 Hz to 1 kHz. The high frequency adjustments are done afterwards. The frequencies used for high frequency adjustments are largely determined by the capacitive vs. resistive characteristics of the scaling circuits. When making a high frequency adjustment, one of two actions is performed:

(a) The actual frequency response of a scaling amplifier or attenuator is physically adjusted.
(b) The digital multimeter's frequency response at cardinal points is stored in its non-volatile memory. The digital multimeter firmware then displays the correct reading at any frequency by interpolating between frequency points.

The second method requires a digital multimeter with an internal frequency counter to determine input frequency.

Resistance converter calibration

Calibration of resistance converters generally consists of zero and gain adjustments. For example, each range of the digital multimeter's resistance function is first corrected for proper zero reading. Then a near-full-scale input is applied to each range, and adjustments are made if necessary.

When calibrating digital multimeters with 4-wire resistance capability, resistance sources should have remote sensing capability (e.g. as provided in the Fluke 5450A resistance calibrator or the Fluke 5700A multifunction calibrator).

In the 2-wire set-up, the sense path and ohms current path are on the same set of terminals. Resistance errors due to the connecting leads can be significant. In general, proper connection of 2-wire resistance calibration for a meter with even 100 mΩ resolution can be a problem (e.g. in the Fluke 5700A this problem has been addressed by incorporating a 2-wire ohms compensation feature).

Separate 2- and 4-wire resistance calibrations could be specified in calibration procedures. Since the 4-wire sense path is the same as the DC voltage signal conditioning path, all that is required is to correct the gain of the current source for resistance converters with current sources. However, second-order effects cause discrepancies in apparent gain when the DC voltage amplifiers are configured to sense either a DC voltage input directly or the voltage across the unknown resistor. Because of this, many of the higher precision digital multimeters require calibration for the 2- and 4-wire resistance.

8.7.3.4 Calibration of current converters

Both AC and DC current converters are calibrated by applying a known current and adjusting for the correct reading. Direct current converters correct for zero and gain; alternating current converters correct for down-scale and full-scale response. Typically, there are no high frequency adjustments for alternating current, primarily because current shunts are not inductive enough to cause significant errors relative to the digital multimeter's specifications.

Closed case calibration

The ability to calibrate a DMM using internally stored offset and gain corrections is known as closed case calibration. A significant number of present day DMMs, particularly bench and laboratory types, have the ability to be calibrated in the closed case mode. All major manufacturers now provide closed case calibration facilities in their top of the range instruments.

There is a common misconception that instruments having a closed case facility do not require calibration using external standards. This is not correct. In order to determine the corrections that are stored internally, a known set of values has to be applied at the input of the instrument. These known values could only be obtained from an external standard.

Closed loop calibration

A system that enables the measurement of an instrument performance, and its automated closed case adjustment, is known as a closed loop calibration system. The main components of a closed loop system are the unit under test, a device to provide the calibration stimuli such as a multifunction calibrator and a computer, interconnected using the IEEE-488 interface bus.

The computer is programmed to address the unit under test through the interface bus and conduct the calibration. It takes the as-found test data, analyses the results and adjusts the unit under test as required. The unit under test of course must be able to be adjusted without manual operator intervention, i.e. having the closed case calibration facility. The majority of closed loop calibrations involve DMMs, since their readings can be transferred easily to a computer for evaluation.

Bibliography

Introductory reading

1. *Calibration: Philosophy in Practice* (1994) Fluke Corporation, USA.
2. Kupferman, S.L. and Hamilton C.A. (1998) Deployment of a compact, transportable, fully automated Josephson voltage standard, National Institute of Standards and Technology paper.
3. Inglis, D., Wood, B., Young, B. and Brown, D. (1999) A modular, portable, quantised Hall resistance standard, *Cal. Lab.*, July–Aug. pp. 27–31.
4. Eicke, W.G. and Cameron, J.M. (1967) Designs for surveillance of the volt maintained by a small group of saturated cells, National Bureau of Standards Technical Note 430, October.
5. Delahye, F. (1992) DC and AC techniques for resistance and impedance measurements, *Metrologia*, 29, pp. 81–93.

Advanced reading

1. Ott, W.E. (1974) A new technique of thermal rms measurement. *IEEE Journal. Solid State Circuits*, Vol. SC-9, pp. 374–380.

9

Uncertainty of measurements

9.1 Introduction

The uncertainty to be assigned to a measurement result has been a widely debated subject from the very beginning of modern measurements. At least two different approaches for determining the total uncertainty had been used in the past, namely the combining of uncertainties arithmetically or in quadrature (root sum of squared quantities). In 1990 the International Committee of Weights and Measures (CIPM) convened a group of experts to formulate a consistent set of rules to compute uncertainties of measurements. The group came out with a set of recommendations known as CIPM recommendations for computation of uncertainties. In 1993, the International Organization for Standardization published the recommendations in the form of a guide. The guide is now widely used and is known as the ISO Guide for Measurement Uncertainties.

Since the publication of the CIPM recommendations, a consensus in the procedures for determining the combined uncertainty has emerged among a majority of the national standards laboratories. This is a positive development for everybody involved in metrology at all levels.

9.2 Basic concepts

The basic objective of performing a measurement is to determine the value of a specific quantity, i.e. the value of a measurand. A measurement therefore begins with an appropriate specification of the measurand, the method of measurement and the measurement procedure.

Unfortunately the value of the measurand cannot be determined exactly due to errors that arise during the measurement process. It is only possible to obtain an estimate for the value of the measurand. This estimate is complete only when it is supplemented by a statement indicating its inexactness, namely the uncertainty.

The terms *error*, *random error* and *systematic error* were used (see Chapter

2) in the past to describe uncertainties and have caused much confusion. If we accept the fact that no measurement can be exact, then these terms can be defined unambiguously. In Chapter 1, *error* was defined as the difference between the result of a measurement and the *true value* of the measurand. Since the *true value* cannot be determined, the so-called *error* cannot be determined either.

The concept of uncertainty relieves us of this problem. The word *uncertainty* means doubt and thus in its broadest sense *uncertainty of measurement* conveys the idea that the measurement result is not exact. The formal definition of the term as given in the current edition of the international vocabulary of metrology (VIM) is:

> *the parameter associated with the result of a measurement that characterizes the dispersion of the values that could reasonably be attributed to the measurand.*

This definition may be contrasted with two older concepts:

– *a measure of the possible error in the estimated value of the measurand as provided by the result of a measurement.*
– *an estimate characterizing the range of values within which the true value of a measurand lies.*

(VIM, first edition, 1984, 3.09)

These two definitions are based on quantities that cannot be determined; the *error* of a result of measurement and the *true value* of a measurand respectively. These definitions are therefore ideals whose practical application is impossible. In the context of the present definitions it is best that the words *error* and *true value* are avoided in test reports and other technical documents.

9.3 Recommendations of the ISO guide

The main recommendations of the guide are summarized here. For fuller details and comprehensive discussion of uncertainty concepts the guide must be consulted.

9.3.1 Types of evaluation

Two methods of evaluation of measurement uncertainties, namely Type A and Type B are recommended in the guide.

9.3.1.1 Type A evaluation

In a Type A evaluation a series of repeated observations is obtained to determine the variance of the measurement result. The positive square root of the variance is referred to as a Type A standard uncertainty (u_A).

9.3.1.2 Type B evaluation

In a Type B evaluation information found in calibration reports, certificates, manufacturer's specifications, etc., are used to estimate a standard deviation known as a Type B standard uncertainty, (u_B).

The designations Type A and Type B are not new words for *random error* and *systematic error*. These are methods of evaluating uncertainty and could equally be applied to both random and systematic effects, e.g. the uncertainty of a correction for a known systematic effect may be obtained by either a Type A or Type B evaluation, as may be the uncertainty due to random effects of a measurement result.

9.3.1.3 Evaluation of Type A uncertainty

The mean value $\bar{q}$ of m independent observations $q_1, q_2, \ldots q_k, \ldots q_m$ is obtained from:

$$\bar{q} = \frac{1}{m}\left(\sum_{k=1}^{m} q_k\right) \tag{9.1}$$

The variance of observations is given by:

$$s^2 = \frac{1}{m-1}\sum_{k=1}^{m}(q_k - \bar{q})^2 \tag{9.2}$$

The variance of the mean $\bar{q}$ is given by

$$s^2(\bar{q}) = \frac{s^2(q)}{m} \tag{9.3}$$

The number of observations m should be sufficiently large so that $\bar{q}$ provides a reliable estimate of the mean value and $s^2(\bar{q})$ provides a reliable estimate of the variance of the mean.

The Type A standard uncertainty $u_A(q)$ is obtained as:

$$u_A(q) = s(\bar{q}) = \frac{s(q)}{\sqrt{m}} \tag{9.4}$$

The number of degrees of freedom v is given by:

$$v = m - c \tag{9.5}$$

where m is the number of observations and c the number of constraints.

9.3.1.4 Evaluation of Type B uncertainty

In a Type B evaluation, the estimated variance $s^2(q)$ or standard deviation $s(q)$ of a quantity q is obtained by judgement using all relevant information on the possible variability of q. The pool of information may include previous measurement data, experience with or general knowledge of the behaviour and properties of relevant materials and instruments, manufacturer's specifications, data provided in calibration and other certificates, and uncertainties assigned to reference data taken from handbooks.

The proper use of the pool of available information for a Type B evaluation of standard uncertainty calls for insight based on experience and general knowledge, but is a skill that can be learned with practice. It should be recognized that a Type B evaluation of standard uncertainty can be as reliable as a Type A evaluation, especially in a measurement situation where a Type A evaluation is based on a comparatively small number of statistically independent observations.

The following guidelines are used for obtaining Type B evaluations:

(a) A stated confidence interval having a stated level of confidence such as 95 or 99 per cent should be converted to a standard uncertainty by treating the quoted uncertainty as if a normal distribution had been used to calculate it and dividing it by the appropriate factor for such a distribution. These factors are 1.960 and 2.576 respectively for the two levels of confidence given.

(b) If a confidence interval with the stated level of confidence and degrees of freedom is given (e.g. 95 or 99 per cent confidence level with 12 degrees of freedom), then to obtain the standard uncertainty, divide the semi-confidence interval by the student's t-value (see Appendix 2 for a table of student's t values) corresponding to the degrees of freedom.

(c) If an expanded uncertainty and k factor are given, divide the expanded uncertainty by the k factor to obtain the standard uncertainty.

(d) Model the quantity by a uniform, rectangular or triangular probability distribution and estimate the lower and upper limits a_- and a_+ for the value of the quantity in question such that the probability that the value lies in the interval a_- to a_+ for all practical purposes is 100 per cent. The best estimate of the mean value of the quantity is then $(a_- + a_+)/2$.

For a rectangular distribution the standard uncertainty is given by:

$$u_\mathrm{B} = \frac{a}{\sqrt{3}} \tag{9.6}$$

where $a = (a_+ - a_-)/2$, i.e. the semi interval.

If the distribution used to model the quantity is triangular rather than rectangular, then:

$$u_\mathrm{B} = \frac{a}{\sqrt{6}} \tag{9.7}$$

The degrees of freedom to be associated with the standard uncertainty obtained from a Type B evaluation is obtained from the following equation:

$$v_i = \frac{1}{2}\left[\frac{\Delta u(\overline{q})}{u(\overline{q})}\right]^{-2} \tag{9.8}$$

where:

$u(\overline{q})$ = standard uncertainty of $\overline{q}$

$\Delta u(\overline{q})$ = uncertainty of standard uncertainty

$\dfrac{\Delta u(\overline{q})}{u(\overline{q})}$ = relative uncertainty of standard uncertainty.

9.3.2 Determination of combined standard uncertainty and effective degrees of freedom

In many instances a measurement result is obtained by the use of an equation that combines different input quantities. The equation represents the relationship between the measurand and physical quantities on which the measurand is dependent. In such cases the uncertainty of the measurand is computed by combining the standard uncertainties of the input quantities as a combined standard uncertainty.

9.3.2.1 Combined standard uncertainty

The combined standard uncertainty of a measurement result y related to input values $x_1, x_2, \ldots x_n$ by the relationship:

$$y = f(x_1, x_2, \ldots, x_n) \tag{9.9}$$

is given by:

$$U_c^2(y) = \left(\frac{\partial f}{\partial x_1}\right)^2 U^2(x_1) + \left(\frac{\partial f}{\partial x_2}\right)^2 U^2(x_2) + \ldots + \left(\frac{\partial f}{\partial x_n}\right)^2 U^2(x_n)$$

$$\tag{9.10}$$

where $U(x_1), U(x_2) \ldots U(x_n)$ are standard uncertainties evaluated as described in the previous section (Type A evaluation or Type B evaluation) for each input quantity. Equation (9.10) is valid only when the input quantities $x_1, x_2, x_3, \ldots x_n$ are independent, i.e. uncorrelated. If the input quantities are correlated then the covariances have to be taken into account. The following equation applies:

$$U_c^2(y) = \sum_{i=1}^{n}\left(\frac{\partial f}{\partial x_i}\right)^2 U^2(x_i) + 2\sum_{i=1}^{n-1}\sum_{j=i+1}^{n}\left(\frac{\partial f}{\partial x_i}\right)\left(\frac{\partial f}{\partial x_j}\right)U(x_i, x_j) \tag{9.11}$$

In this equation the terms $U(x_i, x_j)$ are the covariances of the input parameters x_i and x_j where $j = i + 1$. Equation (9.11) is known as the *law of propagation of uncertainties*. When the input quantities are independent (not correlated) the second term in equation (9.11) is zero and equation (9.10) results.

9.3.2.2 Effective degrees of freedom

Welch–Satterthwaite formula

The effective degrees of freedom of the combined standard uncertainty is obtained from the Welch–Satterthwaite formula:

$$v_{\text{eff}} = \frac{u_c^4(y)}{\sum\limits_{i=1}^{n} \dfrac{u_i^4(y)}{v_i}} \tag{9.12}$$

where:

$$U_i(y) = \left(\frac{\partial f}{\partial x_i}\right) U(x_i)$$

v_i = degrees of freedom of input uncertainty $U_i(y)$.

The value of v_{eff} resulting from equation (9.12) may not be an integer. In this case v_{eff} is obtained by rounding down to the next lower integer.

9.3.3 Expanded uncertainty

The expanded uncertainty is analogous to the confidence interval of a measurement result. The expanded uncertainty **U** is obtained by multiplying the combined standard uncertainty $u_c(y)$ by a coverage factor **k**:

$$\mathbf{U} = \mathbf{k} \cdot U_c(y) \tag{9.13}$$

The result of a measurement is then expressed as $y \pm \mathbf{U}$ at a confidence level of 95 per cent or 99 per cent. The value of the coverage factor **k** is chosen on the basis of the desired level of confidence to be associated with the interval $y - \mathbf{U}$ to $y + \mathbf{U}$. In general **k** is chosen in the range 2 to 3.

The method of choosing **k** using the effective degrees of freedom v_{eff} and student's t distribution is explained in the examples.

9.4 Examples of uncertainty calculations

9.4.1 Example 1 – Determination of the uncertainty of the value of a test mass

9.4.1.1 Experimental data

The value of a stainless steel mass of nominal value 1 kilogram is determined by carrying out double substitution weighing in a single pan balance, using a reference standard mass of known value and uncertainty. The weighing operation is repeated ten times and the results are given in Table 9.1.

Table 9.1 Weighing results for calibration of a test mass

No.	Value of test mass g
1	1000.143
2	1000.144
3	1000.144
4	1000.146
5	1000.146
6	1000.146
7	1000.144
8	1000.143
9	1000.145
10	1000.145

Mean value of test mass = 1000.1446 g
Standard deviation = 0.0011 g
Standard deviation of the mean = $0.0011/\sqrt{10} = 0.00035$ g

9.4.1.2 Estimation of combined standard uncertainty $U_c(y)$

The contributory uncertainties arise from:

(a) Variability of the weighing procedure quantified by the standard deviation of the mean given above.
(b) The uncertainty of the mass standard used. This is taken from the calibration certificate of the standard mass and is given as 0.005 g at 95 per cent confidence level with 14 degrees of freedom.
(c) The uncertainty arising from the resolution of the digital display of the balance used for the comparison, 0.001 g. This is treated as having a rectangular probability distribution and the standard uncertainty is computed by dividing the half interval by $\sqrt{3}$, i.e. $0005/\sqrt{3}$ g.

These are summarized in Table 9.2.

Table 9.2 Uncertainty budget of test mass

Source of uncertainty	Type of evaluation	Standard uncertainty	Degrees of freedom
Variability of observations	A	0.000 35 g	9
Reference mass standard	B	0.0025 g	14
Resolution of the digital display	B	0.000 29 g	infinity

The estimation of the combined standard uncertainty of the mean value is given below:

Combined standard uncertainty,

$$U_c(y) = (0.000\,35^2 + 0.002\,5^2 + 0.000\,29^2)^{1/2}$$

$$= 0.002\,54 \text{ g}$$

$$= 0.002 \text{ g (rounded off to one significant figure)}$$

9.4.1.3 Effective degrees of freedom

The effective degrees of freedom is computed using equation (9.12):

$$v_{\text{eff}} = \frac{0.002^4}{\dfrac{0.000\,35^4}{9} + \dfrac{0.0025^4}{14} + \dfrac{0.000\,29^4}{\infty}} = 9.6$$

This is rounded down to 9 degrees of freedom.

9.4.1.5 Expanded uncertainty

To determine expanded uncertainty, it is necessary to choose a coverage factor, **k**. The coverage factor is taken from the t tables, corresponding to 9 degrees of freedom and 95 per cent confidence level, $t_{95,9} = 2.26$:

Expanded uncertainty, $\mathbf{U} = 2.26 \times 0.002$

$$= 0.0045 \text{ g}$$

9.4.1.4 Reporting of the result

The result is reported as:

Value of the test mass = 1000.145 g

Expanded uncertainty = ±0.005 g with $k = 2.26$ at 95 per cent
confidence level and 9 degrees of freedom

or

The value of the test mass is 1000.145 g ± 0.005 g with $k = 2.26$ at 95 per cent confidence level and 9 degrees of freedom.

9.4.2 Example 2 – Determination of the value of a test resistance

A test resistance is calibrated using a direct reading digital multimeter and resistance standard. The test resistance is measured ten times using the digital multimeter and the reading ρ_x is noted down. The standard resistance is then measured ten times using the same multimeter and the corresponding reading ρ_s is noted down.

The value of the test resistance (R_x) is calculated using the equation:

$$R_X = R_S \frac{\overline{\rho}_X}{\overline{\rho}_S} \tag{9.14}$$

where R_S is the calibrated value of the standard resistance, $\overline{\rho}_X$ the mean of ρ_X and $\overline{\rho}_S$ the mean of ρ_S.

9.4.2.1 Experimental data

The experimental data are given below:

$$\overline{\rho}_x = 100.051 \ \Omega$$

Standard deviation = 18 ppm

$$\overline{\rho}_s = 100.001 \ \Omega$$

Standard deviation = 19 ppm

$$R_s = 99.999 \ \Omega$$

Expanded uncertainty = 20 ppm at $k = 2$ with 9 degrees of freedom

To calculate the value of the test resistance and its uncertainty, we proceed as follows.

9.4.2.2 Value of test resistance

Using equation (9.14):

$$R_x = 99.999 \times \frac{100.051}{100.001} = 100.0489 \ \Omega$$

9.4.2.3 Calculation of uncertainty

The main uncertainty components are:

(a) Transfer uncertainty of the digital multimeter.
(b) Calibration uncertainty of the standard resistor.

Transfer uncertainty of the digital multimeter (Type A)

Since both test and standard resistances are measured using the same digital multimeter under similar conditions it is possible to determine the transfer uncertainty as a pooled standard deviation of the two measurement standard deviations.

$$\text{Pooled } s = \sqrt{\frac{18^2 + 19^2}{2}}$$

$$= 18.5 \text{ ppm with 18 degrees of freedom } (18 = 10 \times 2 - 2)$$

$$\text{Standard deviation of the mean} = U_{tr} = \frac{18.5}{\sqrt{10}} = 5.8 \text{ ppm}$$

Standard uncertainty = U_{tr} = 5.8 ppm with 18 degrees of freedom

Calibration uncertainty of the standard resistance (Type B)

This is given as 20 ppm at $k = 2$:

$$\text{Standard uncertainty, } U_{R_x} = \frac{20}{2} = 10 \text{ ppm}$$

Combined standard uncertainty

The combined standard uncertainty is computed as:

$$U_{R_x} = \sqrt{5.8^2 + 10^2} = 11.56 \text{ ppm}$$

Effective degrees of freedom

The effective degrees of freedom is computed using equation (9.12) as:

$$v_{\text{eff}} = \frac{11.56^4}{\dfrac{5.8^4}{18} + \dfrac{10^4}{9}} = 15.21 \text{ round down to } 15$$

Expanded uncertainty

The t value for 95 per cent confidence level and 15 degrees of freedom is 2.13. Therefore, $k = 2.13$.

The expanded uncertainty (**U**) is given by:

$$\textbf{U} = 2.13 \times 11.56 = 24.56 \text{ ppm round off to 25 ppm.}$$

$$25 \text{ ppm} = \frac{25 \times 100.0489}{10^6} = 0.0025 \ \Omega$$

Reporting of results

The value of the test resistor = 100.049 Ω ± 0.002 Ω.

Bibliography

Introductory reading

1. Campion, P.J., Burns, J.E. and Williams, A. (1973) *A code of practice for the detailed statement of accuracy*, National Physical Laboratory, UK.
2. Eisenhart, C. (1963) Realistic evaluation of the precision and accuracy of instrument calibration systems, *Journal of Research National Bureau of Standards*, NBS 67C (Eng. and Instr.), No. 2, 161–187.
3. *Guide to the Expression of Uncertainty in Measurement* (1995). International Organization for Standardization. Corrected and reprinted edition.
4. *International vocabulary of basic and general terms in metrology*, (1993). Second edition, International Organization for Standardization.

Advanced reading

1. Philips, S.D. and Eberhardt, K.R. (1997) Guidelines for expressing the uncertainty of measurement results containing uncorrected bias, *Journal of Research of the National Institute of Standards and Technology*, 102, 577–585.
2. Bich, W. (1996) Simple formula for the propagation of variances and covariances, *Metrologia*, 33, 181–183.
3. Welch, B.L. (1936) *Journal of Research of Statistical Society*, Suppl., 3, 29–48.
4. Satterthwaite, F.E. (1941) *Psychometrika*. Ch. 6, 309–316.
5. Jeffreys, H. (1983) *Theory of Probability*, third edition, Oxford University Press, Oxford.

Appendix 1 _____

Example calibration report

CALIBRATION REPORT

Report No:	Date of Calibration:
Date of Report:	Due Date:

Item Calibrated:

Description: Pressure transmitter
Manufacturer: Endress & Hauser
Model: 335 PT
Serial No: 2134

Reference Standard(s) & Traceability

Description: Pressure calibrator
Manufacturer: Fluke Mfg Co.
Model: 852
Serial No: 3456 PC
Traceability: Fluke Laboratory, USA, Report No. 345 of 23 April 2000

Test Conditions:

Temperature: Ambient temperature (approx. 30°C)
Humidity: Ambient humidity (80–90% RH)
Ambient Pressure: Atmospheric

and any other relevant information pertaining to test conditions can be given here.

Procedure:

Reference the document where the calibration or test procedure is written down.

E.g. document LB/01-1997 – Procedure for calibration of pressure sensors.

Method of Calibration:

Give a brief description of the method of calibration here.

E.g. the pressure sensor was calibrated *in situ* by applying static pressure using a compressed air supply at ambient temperature and pressure.

Results:

Insert a statement such as:

Calibration results are given below. The measured value is the mean value obtained from _____ runs. The correction should be added to the measured values to obtain corrected values.

Nominal	Measured value	Reference std. value	Correction	Uncertainty of measured value
kPa	kPa	kPa	kPa	kPa

Uncertainty:

The uncertainty given above is the expanded uncertainty of the measured value evaluated as the root sum square of Type A and Type B uncertainties with a coverage factor k = 2 (or at a confidence level of 95 per cent).

Calibrated by,   Checked by,

_____   _____
Name & signature of   Name & signature of
calibration officer   quality assurance officer

Approved by,

_____
Name & signature of
authorized signatory
(Manager, Laboratory)

Appendix 2

t table

Value of $t_p(v)$ from the t distribution for degrees of freedom v that defines an interval $-t_p(v)$ to $+t_p(v)$ that encompasses the fraction p of the distribution

Degrees of freedom v	Fraction p in per cent					
	68.27[a]	90	95	95.45[a]	99	99.73[a]
1	1.84	6.31	12.71	13.97	63.66	235.80
2	1.32	2.92	4.30	4.53	9.92	19.21
3	1.20	2.35	3.18	3.31	5.84	9.22
4	1.14	2.13	2.78	2.87	4.60	6.62
5	1.11	2.02	2.57	2.65	4.03	5.51
6	1.09	1.94	2.45	2.52	3.71	4.90
7	1.08	1.89	2.36	2.43	3.50	4.53
8	1.07	1.86	2.31	2.37	3.36	4.28
9	1.06	1.83	2.26	2.32	3.25	4.09
10	1.05	1.81	2.23	2.28	3.17	3.96
11	1.05	1.80	2.20	2.25	3.11	3.85
12	1.04	1.78	2.18	2.23	3.05	3.76
13	1.04	1.77	2.16	2.21	3.01	3.69
14	1.04	1.76	2.14	2.20	2.98	3.64
15	1.03	1.75	2.13	2.18	2.95	3.59
16	1.03	1.75	2.12	2.17	2.92	3.54
17	1.03	1.74	2.11	2.16	2.90	3.51
18	1.03	1.73	2.10	2.15	2.88	3.48
19	1.03	1.73	2.09	2.14	2.86	3.45
20	1.03	1.72	2.09	2.13	2.85	3.42
25	1.02	1.71	2.06	2.11	2.79	3.33
30	1.02	1.70	2.04	2.09	2.75	3.27
35	1.01	1.70	2.03	2.07	2.72	3.23
40	1.01	1.68	2.02	2.06	2.70	3.20
45	1.01	1.68	2.01	2.06	2.69	3.18
50	1.01	1.68	2.01	2.05	2.68	3.16
100	1.005	1.660	1.984	2.025	2.626	3.077
∞	1.000	1.645	1.960	2.000	2.576	3.000

[a]For a quantity z described by a normal distribution with expectation μ and standard deviation σ, the interval $\mu z \pm k\sigma$ encompasses $p = 68.27$, 95.45 and 99.73 per cent of the distribution for $k = 1$, 2 and 3, respectively.

Appendix 3 _____
Useful addresses

1. American Technical Publishers Ltd, 27–29 Knoll Place, Wilbury Way, Hitchin, Herts SG4 0SX
   Tel: 01462 437933   Fax: 01462 433678
   E-mail: atpltd@compuserve.com
   Website: www.ameritech.co.uk
2. The British Library, Technology and Business Services, 96 Euston Road, London NW1 2DB
   Tel: 020 741294/96   Fax: 020 7412 7495
   Website: www.bl.uk\services\stb
3. BSI Standards, 389 Chiswick High Road, London W4 4AL
   Tel: 020 8996 9000   Fax: 020 8966 7400
   Website: www.bsi.org.uk
4. Institute of Measurement and Control, 87 Gower Street, London WC1E 6AA
   Tel: 020 7387 4949   Fax: 020 7388 8431
   Website: www.instmc.org.uk
5. Institute of Petroleum, 61 New Cavendish Street, London W1M 8AR
   Tel: 020 7467 7100
   Website: www.petroleum.co.uk
6. Institution of Electrical Engineers, Savoy Place, London WC2R 0BL
   Tel: 020 7240 1871   Fax: 020 7240 7735
   Website: www.iee.org.uk
7. Instrument Society of America (ISA) Standards are available from American Technical Publishers (see 1 above).
8. National Association of Testing Authorities (NATA)
   71–73 Flemington Road, North Melbourne, VIC 3051
   Australia
   Tel: 613 9329 1633   Fax: 613 9326 5148
   Website: www.nata.asn.au
9. National Physical Laboratory, Queens Road, Teddington, Middlesex TW1 0LW
   Tel: 020 8977 3222   Fax: 020 8943 6458
   Website: www.npl.co.uk
10. National Conference of Standards Laboratories
    1800, 30th Street, Suite 305B, Boulder, CO 80301
    Tel 303 440 3339

11. National Institute of Standards and Technology (NIST)
    Gaithesburg, MD 20899-0001, USA
12. National Measurement Laboratory
    PO Box 218
    Lindfield, NSW 2070
    Australia
13. SIRA Test and Certification Ltd, South Hill, Chislehurst, Kent BR7 5EH
    Tel: 020 8467 2636   Fax: 020 8467 6515
    E-mail: siraltdinfo@sira.co.uk
    Website: www.sira.co.uk
14. UKAS, Queens Road, Teddington, Middlesex TW1 1 0LW
    Tel: 020 8943 7140   Fax: 020 8943 7134

Index